Flashing Blades over the Sea

The Development & History of Helicopters in the Royal Navy

By

Lt. Cdr. J. M. Milne RN

Foreword By

HRH The Princ

£1.80

An injured seaman being transferred from a tanker in Mid-Atlantic to HMS ARK ROYAL by a Sea King HAS 2 of 824 Squadron. The photograph was taken by a shepherding aircraft.

Buckingham Palace

Foreword by
His Royal Highness The Prince of Wales KG

As the title suggests, this book is concerned with the history of the helicopter and its adoption by the Fleet Air Arm. With the first helicopter deck landing taking place in 1943, the history covers less than forty years. Yet, as the book reveals, the art has developed in that short time at an exciting pace.

From the early days of the Dragonfly through to the Lynx and beyond, it is an interesting story to tell, including such historical milestones as the Malayan Emergency, Suez, Indonesian Confrontation, Cyprus and the endless Search and Rescue missions. Thanks to the imagination and initiative shown over the last forty years the Fleet Air Arm now has helicopters which keep the Royal Navy well ahead in the field of Anti Submarine Warfare and provide a versatility which would have been unthinkable a few years ago. The author is the first to attempt a history of this subject and he has done so in a comprehensive manner. As one associated all too briefly with this fascinating tale, I commend the book to all helicopter enthusiasts and to all members, past and present, of the Naval Air Squadrons described within.

Charles.

INTRODUCTION

The past few years have seen a major change in the Fleet Air Arm. Fixed-wing flying ceased with the paying-off of H.M.S. ARK ROYAL, whilst the arrival of H.M.S. INVINCIBLE was awaited. Helicopter operations by the Royal Navy have come into their own.

New helicopters such as the Lynx H.A.S. Mk 2 are flying in the Fleet and others are on the drawing board. How has the Fleet Air Arm arrived at the Lynx and how has the Royal Navy arrived at the present state of helicopter warfare in the Anti-Submarine, Anti-Surface and Commando roles?

Having served with the Fleet Air Arm for 20 of the 36 years of the development of the tactics and techniques used with the helicopter, I felt it was time to set down the history of the aircraft as used and developed by the Royal Navy. My aim has been to produce a book which will serve not only as a reference for the aviation enthusiast but also as a readable history for the more general public who have an interest in the affairs of their Armed Forces.

A proportion of the profits of this book are going to the Fleet Air Arm Museum Appeal Fund to assist the Museum at RNAS Yeovilton, in its task of preserving the history of the Fleet Air Arm.

Finally, this work is dedicated to the Fleet Air Arm helicopter aircrew, many of whom I knew personally, that have lost their lives in the service of the Royal Navy.

Signed

James M Milne.

Lt Cdr R.N.

CHAPTER 1

The Beginnings

In order to set the scene for this book it is first necessary to recap on the known history of the helicopter.

Leonardo da Vinci is credited with the first design of a helicopter in his famous sketch produced around 1483. The next major events in the calendar are in the 18th century when the Russian and French Academies of Science had working model helicopters demonstrated to them by Mikhail Lomonosov and Launoy and Bienvenu respectively. Around 1863 the name helicopter was coined from the Greek words, *Helicos* (spiral) and *Pteron* (wing), by Vicomte Gustave de Ponton d'Amecourt.

R4B, R6 and S51 [Dragonfly] in formation.

In 1907 experiments and experience finally produced the first helicopter carrying a pilot to lift off from the ground. On the 24 August, at Douai in France, a machine designed and built by Louis and Jacques Breguet, names later to become famous in the French aircraft industry, lifted 2 inches off the ground.

In 1909 a name which would eventually be closely linked

with helicopters operated by the Fleet Air Arm appeared on the scene, that of Igor Sikorsky. He designed and built a machine with twin co-axial rotors in Russia. Unfortunately the design was not successful and after more failures he gave up and turned to fixedwing aircraft.

World War I slowed down the rate of development of helicopters and no new designs emerged. Between 1920 and 1936 war technology assisted the Spaniard Pescara, the Frenchman Louis Brequet, the Russian de Bothezat (in America) and the German Foche to make great steps forward and it was during this period that the Admiralty started to take an interest in rotary wing flying. Not in helicopters, but in autogyros.

The Chief Pilot for the Cievra Autogyro Company, Reginald Brie had made several landings and take-offs from the Italian cruiser FIUME, anchored and underway (up to 24 knots) off La Spezia in 1935, and the Admiralty was so impressed that they asked for the trials to be repeated. The Cievra Autogyro Company agreed, providing a definite order would be forthcoming from the Admiralty. The Admiralty allowed itself to be pressurised and ordered two C30 autogyros. These were 2 seater aircraft with a 3 bladed 'direct control' rotor of 37 feet diameter powered by a 140 h.p. Armstrong Siddeley Genet engine.

The autogyros were delivered to R.A.F. Gosport and after some initial engineering problems successful Deck Landing trials were carried out onboard the aircraft carrier H.M.S. FURIOUS in the English Channel.

As on ships, apart from aircraft carriers, deck space was limited, it was obviously desirable to achieve a complete elimination of the already short autogyro take-off run. This was achieved in 1938 by the C40 powered by a 180 h.p. Salmson engine, and the Admiralty ordered 5 of these machines. Two C40's were delivered just prior to the outbreak of the Second World War, the other three going to France with the British Expeditionary Force for Army observation work (flown by RAF pilots). The intention for the two remaining C40's was as spotters in the anti-submarine and night shadowing roles. No evidence exists however as to what results, if any, they achieved.

Whilst the Autogyro story was developing in the Royal Navy the Ministry of Supply was carefully watching developments in Germany. On 26 June 1936 Doctor Heinrich K.J. Foche saw his FW61 helicopter take to the air. By 1939 the FW61 had established a world altitude record of 11,243 feet and a distance record of 141 miles. The German Navy saw the potential of this type of aircraft and put out a requirement for a small helicopter capable of operating from small ships or submarines. The British Ministry of Supply had also seen this potential and tried to purchase an FW61 for flight trials. The Germans however, well aware of the value of their FW61, quoted excessive prices and hedged over delivery dates so well, that no FW61 ever arrived in Britain.

It is worthy of note that in Britain at this time there existed a design team of C.G.Pullin (Chief Designer with Cievra) and the Glasgow engineering firm of C & J Weir which had produced a helicopter designated the W6. The W6, which first flew on 29 Oct 1939, was of advanced design but unfortunately had to be abandoned in July 1940 owing to war-time constraints. If development of the W6 had been allowed to continue it is possible that the Royal Navy would have been operating helicopters two or three years before the actual operational date.

In 1939, Igor Sikorsky, who had fled from Russia to America and returned to his first love — helicopters, had successfully flown his VS-300 on 14 September. This first flight was a 'tethered' flight, the first free flight not occurring until 13 May the following year. The U.S. Navy at this stage could see no future in helicopters and the U.S. Army Air Corps was tasked to watch developments and procure aircraft for trials.

In 1941 the British Director of Naval Development and Production, Captain Caspar John, (the late Admiral of the Fleet, Sir Caspar John), sent Reginald Brie, by now a Wing Commander in the R.A.F., on a mission to the United States to interest the American services, (in particular the U.S. Navy,) in the use of autogyros in the anti-submarine role.

Early in 1942, just prior to his demonstrations of the autogyro to the Americans onboard the Merchant Aircraft Carrier H.M.S. RANGER and the Merchantman EMPIRE MERSEY, Wing Commander Brie was witness to a very impressive demonstration of the VS300. Also, in January 1942 the first flight of the XR-4 had taken place, this aircraft being a direct result of the interest that the Army Air Corps had shown in helicopters. (The Army Air Corps having awarded Sikorsky a contract to develop the potential that the VS300 had shown.) Wing Commander Brie must have realised at this time that the days of the autogyros were numbered.

The British Air Commission in Washington had been

following the XR-4 development and decided, in conjunction with the United States Coastguard, that one of the major roles for helicopters could be in A.S.W. (Anti-Submarine Warfare) convoy protection. The case for the autogyro was abandoned as a lost cause. The Coastguard ordered 50 R-4s and the British 25, although 10 of the latter were for R.A.F. use. On 1 January 1943 Sikorsky set up a production line for R-4s at Stratford, Connetticutt.

A joint Board for the Evaluation of the Helicopter in Anti-Submarine Warfare was formed in early 1943 — under the direction of Admiral E. King, U.S. Navy. Members included Officers from the U.S. Coastguard and the Royal Navy. One of the first tasks of the Board was to evaluate the feasibility of operating helicopters at sea in all conditions, preferably from platforms rather than aircraft carrier decks. The merchantman EMPIRE MERSEY, which was fitted with the platform for the

Sikorsky R4 during Deck Landing Trials onboard USS CHARGER.

Autogyro trials, was sunk in 1943; so the British merchantman DAGHESTAN was requisitioned, and a platform similar to that of the EMPIRE MERSEY fitted. Dimensions were approximately 90 feet by 40 feet. An XR-4 had landed on the U.S.S. BUNKER HILL early in 1943 and that was believed to be the first landing by a true helicopter on a ship at sea.

An R4 making a landing onboard the SS DAGHESTAN during a World War 2 Atlantic convoy.

The trials conducted aboard the DAGHESTAN off the American North Eastern seaboard were successful in that they proved the feasibility of open sea operation. However, operational trials were restricted solely to observation duties as the two XR-4 aircraft fitted with 180 h.p. Warner Super Scarab engines had no weapon carrying capability.

By the end of 1943 the Joint Board was convinced by the XR-4 trials and ordered two R-4s to be embarked in the DAGHESTAN for a 50 ship convoy crossing of the Atlantic. Included in the flight trials crew was one Royal Navy pilot, two RAF pilots and one U.S. Coastguard pilot. An extremely rough Atlantic crossing prevented the trial being carried out fully but the Coastguard pilot did manage one take-off and landing in mid-Atlantic. The trial, begun from New York on 2 January 1944, proved inconclusive and the Joint Board recommended that the trials should continue, although in its present form the helicopter was not suitable as an anti-submarine weapon. The Royal Navy had, as a result of the earlier promising trials, ordered a quantity of an improved R-4 — the R.5. Accepting the

Board's recommendations, the Royal Navy cancelled the R-5 order, preferring to save the money for more powerful aircraft which were then being designed.

A total of 240 R-4Bs were ordered by the Royal Navy, the R-4B being a more powerfully engined R-4. Due to the end of hostilities only 45 were actually delivered; under the Lend-Lease scheme. The majority went to the Fleet Air Arm and the rest to the R.A.F.

Whilst all the trials were going on in America and at sea, the R-4Bs already delivered to Britain were being put to good use. A trials Squadron, 771 Squadron, was the first Fleet Air Arm unti to received the R-4B on strength, in September 1945. The R-4Bs were based at Portland and were given the very apt name of HOVERFLY 1.

At an early stage the Admiralty had requested training for R.N. pilots from the U.S. Coastguard who had been operating R-4Bs. The U.S. Coastguard agreed and No. 1 Helicopter (Instructors) Course started at the U.S. Coastguard Station, Floyd Bennett Field near New York in March 1944. The course consisted of six Royal Navy pilots, two R.A.F. pilots and one Army pilot. The requirement was for all pilots to already have 500 hours solo on fixed-wing aircraft. A seventh Naval pilot with only 60 hours fixed-wing experience joined the course at a later stage and his results did not bear out the original requirement, in that he adapted as well, if not better, than some.

An R4B makes a simulated Deck Landing onboard a 'Queen Mary' trailer at an Air Show at RNAS GOSPORT.

As there was a scarcity of R-4Bs during this time, the Royal Navy agreed that some of its aircraft, instead of being shipped across the Atlantic, should be retained at Floyd Bennett Field and used by both R.N. and Coastguard pupils. Lieutenant Commander Peat, the first Royal Navy helicopter pilot was the liaison officer between the U.S. Coastguards and the Admiralty during this time and he spent a considerable amount of time commuting between the U.S.A. and Britain.

Training on these courses followed a normal pattern. A series of ground lectures, followed by an aerial demonstration, ending up with the pupil in the right hand seat, taking controls. Initial sorties were in R-4Bs with a normal tricycle undercarriage and at a later stage pupils graduated to the few R-4Bs fitted with flotation gear for water landings. A 'rolling deck', (a very realistic mock-up of a ship's stern) also formed part of the course. One pilot, Lieutenant White, had the dubious honour of sliding off this deck, only managing to remain airborne by the skin of his teeth and to move off rearwards towards safety. During this first course it was discovered that the ability of the helicopter to remain stationary over a fixed point made the 'spotting' of sun worshippers on top of the large mansions around New York a relatively easy business. This practice was stopped after an irate householder complained that a helicopter hovering closer than normal had dislodged all the soot in his chimneys.

It is of interest that No.1 Course were informed that after

An R4B provides an alternative for the NAAFI girl at RNAS WORTHY DOWN.

training they would be dispatched to Britain and used for "spotting" for the invasion over the beaches. This rumour, as that is all it was, was not too well received as the R-4B did have the unfortunate quirk of dropping out of the sky every now and then due to engine failures. The complete course totalled 70 hours of flying and lasted four months.

In the event, helicopters were not used for spotting on the beaches and in Britain there was a lull in helicopter development in an operational sense. The R-4B's in 771 Squadron were used as targets for the calibration of ships radar sets, for photography, for communication duties and training a small number of pilots to complement those being trained in the United States.

The U.S. Coastguard had used R-4B's in rescues by this time and thoughts in the Royal Navy were turning towards the first true operational role for the helicopter in the Fleet Air Arm — Air Sea Rescue.

CHAPTER 2

First Operational Use

By 1946 it was apparent that the helicopter was here to stay and had a potential in both the military and civil aviation fields.

Sikorsky in America had viewed the maiden flight of his most recent development — the S-51, in March 1946. In Britain no design team existed which was capable of producing a helicopter as advanced as the S-51. With both the Services and Civil Aviation showing an interest the Westland Aircraft Company at Yeovil in Somerset obtained a licence to manufacture the S-51 in Britain, using a British powerplant, in December 1946. The relationship between Sikorsky, Westlands and the Fleet Air Arm had begun.

Until Westlands began to produce S-51's the Royal Navy had only its R-4B's to build up a nucleus of pilots and expertise. So the Royal Navy concentrated on obtaining data with a view to the future. In September 1946 Lieutenant Alan Bristow (now owner of Bristow Helicopters Ltd) landed an R-4B on board H.M.S. HELMSDALE, a River Class frigate, at the start of a series of trials connected with anti-submarine warfare. (This landing was the first by a Royal Navy pilot on board a small Royal Navy ship.) Results were inconclusive, but it was now obvious that a helicopter with an improved weight carrying

A float fitted R4B operating from Portland Harbour.

capacity would become a viable A.S.W. machine. Such aircraft were on the drawing board in America but unfortunately the emerging S-51 did not have the required performance. However the Royal Navy was looking for a replacement for the ageing air sea rescue aircraft, the Sea Otter. The S-51 met this requirement.

During 1947 Westlands imported six S-51's for development work. As already stated these were not solely for military development as British European Airways were increasingly showing an interest.

Meanwhile the Royal Navy foreseeing an increase in the number of pilots required, formed its first all-helicopter Squadron, 705 Squadron, at RNAS Gosport on 7 May 1947, with R-4B's transferred from 771 Squadron. Besides training pilots, which was becoming too expensive to carry out in America, 705 Squadron was tasked with undertaking trials and providing aircraft for radar alignment, communication flights, etc.

On 1 February 1947 Lieutenant Ken Reed carried out the first landing by a helicopter on board a major Royal Navy ship. H.M.S. VANGUARD, off Portland whilst the battleship was on its way to South Africa with the Royal Family on board.

Lieutenant K Reed RN makes the first landing onboard a major British warship __ HMS VANGUARD, in an R4B of 705 Squadron.

Dragonfly Mk 1 makes a landing on the cluttered deck of a battleship.

Back at Westlands, the task of 'anglicising' the S-51 was taking longer than first envisaged due to the chronic dollar shortage which prevented the importation of some aircraft parts from America.

The first of 25 R-6's (HOVERFLY 2) had been delivered to Britain during the late summer of 1945 and 15 had been earmarked for Fleet Air Arm use. In fact only 4 passed into Naval service with 705 Squadron at RNAS Gosport and they never proved a successful successor to the R-4B. It seems likely that the first R-6 was received in service around mid 1947 and ended their service in 1950 after an inauspicious career. Why these aircraft were ordered in the first place is not clear.

Finally, on 5 October 1948, after a very frustrating eighteen months for the Westland engineers the first Westland built S-51 flew. Two further prototypes followed within six months. The name DRAGONFLY was adopted for the new aircraft. The Royal Navy impatient to get its hands on the new aircraft and to get trials underway, hired one of the six original imported S-51's. It was allocated to 705 Squadron and used for trials on board the aircraft carrier H.M.S. VENGEANCE in early 1949. The trials were short lived but successful and the aircraft was

Deck Landing Trials onboard RFA FORT DUSQUESNE with a Dragonfly Mk 1.

handed back to Westlands in May 1949.

In the sidelines in 1949 the Bristol Aircraft Corporation were starting work on an aircraft to meet a Ministry of Supply Specification, E4/47. This aircraft was designated the Type 173 and appears in a later chapter.

The first production model of the Westland Dragonfly flew on 22 June 1949 and by June 1950 705 Squadron had taken delivery of 12 Dragonflies, designated HR1. The first delivery flight took place on 13 January 1950. The last R-4B flew on 26 June 1950 when it gave a display at the Southampton Greyhound Stadium.

The Dragonfly HR1 differed from later Dragonfly marks in that its blades were of composite construction with a tubular metal spar, wooden leading edge and fabric covering. Power was supplied by an Alvis Leonides Mk 50. 705 Squadron was now given the additional task of training pilots who would subsequently command air sea rescue units either on board aircraft carriers or at Naval Air Stations ashore.

On 10 November 1949, across the Atlantic, the first flight of an aircraft which would have a profound effect on the Fleet Air Arm took place — the Sikorsky S-55 took to the air. This first development aircraft differed from all other S-55's in that the rear cabin ended at a vertical wall which ran right up to the tail. The distinctive tail fillet was fitted to all other aircraft.

A perfect netting pickup by a Dragonfly Mk 3 from HMS CENTAUR off Malta.

On 12 January 1951 the first Dragonfly Air Sea Rescue Unit embarked on board H.M.S. INDOMITABLE. Also in January of that year a Dragonfly was embarked in RFA FORT DUEQUESNE, which had been fitted with a small landing platform, in order to carry out more trials designed to find out whether helicopters flying from platforms could be successful in all weathers. The worst possible weather was found for the trial and 180 successful landings and take-offs were achieved.

H.M.S. INDOMITABLE was the Flag Ship of the Commander-in-Chief, Home Fleet and saw no active service, but the second Dragonfly Air Sea Rescue Unit embarked in H.M.S. GLORY and sailed for Korea relieving H.M.S. THESEUS in April 1951. The GLORY Dragonfly proved its worth in rescuing 4 pilots from the sea and another 4 from behind enemy lines during the carriers time in Korean waters.

There had been no Dragonfly available for H.M.S. THESEUS before she sailed from Britain and one was borrowed from the Americans on station off Korea. H.M.S. OCEAN relieved H.M.S. GLORY and by this time Dragonflies were firmly established as the prime day time Air Sea Rescue method on board British aircraft carriers. Shore stations were now being supplied with their Air Sea Rescue Dragonflies and a total of fifteen Air Stations and nine carriers operated Flights during the 1950's.

Only 705 Squadron operated the Dragonfly HR1, all other units were equipped with HR3's. The HR3 had all metal rotor blades and hydraulic servo controls. A further mark, the HR5 was introduced in 1957 and involved modifying existing HR1's and HR3's by giving them an increase in engine power and bringing instrumentation up to date. Around 25 HR1 and HR3 models were modified at Royal Naval Aircraft Yard Donibristle, Fife between 1957 and 1960. Further programmed modifications to convert 24 Dragonflies into HC7's (WIDGEON) in 1958 were dropped as a result of a Government economy drive.

Dragonflies proved themselves throughout the Fleet during the 1950's serving, not only with 705 Squadron and Air Sea Rescue Units in the training and rescue roles, but also with 701 Squadron (in the Communications role), 727 Squadron and Britannia Flight (Air Experience Flights), 728 Squadron (Fleet Requirements in Malta at Royal Naval Air Station Hal Far), 771 Squadron (Fleet Requirements at Portland), and aboard H.M.S. VIDAL the first survey ship with a helicopter flight. A Dragonfly relieved a none too successful Hiller HTE-2 on H.M.S. VIDAL. The last Dragonfly in service was finally withdrawn from

Netting practice onboard HMS BULWARK by a Dragonfly Mk 3.

Britannia Flight at Dartmouth in June 1967 giving the type a total of 17 years service to the Fleet Air Arm, a success story indeed for a very early helicopter.

Dragonflies took part in the first major Search and Rescue operation by Fleet Air Arm helicopters in April 1951 when the submarine H.M.S. AFFRAY sank. Two Dragonflies unsuccessfully tried to rescue Captain Carlsen and his Mate from the FLYING ENTERPRISE adrift in the South Western Approaches in January 1952. One aircraft went unserviceable before the actual rescue attempt was made, the second aircraft pressed on after refuelling at the Royal Naval Air Station Culdrose, but was forced back by weather. This attempt, although unsuccessful, did demonstrate that the Dragonfly was capable of flying in conditions thought at that time to be impossible.

Finally, the rescue which brought helicopters truly onto the European and International scene occurred in February 1953 when the sea breeched the sea defences of Holland. All of 705 Squadron Dragonflies were committed to the two week rescue operation which followed. Six hundred people were rescued by 705 Squadron and other Dragonfly operators. The Squadron had the honour of flying Queen Juliana and Prince Bernhardt over the flood stricken area during the relief period. Unfortunately 705 Squadron lost one of its Dragonflies during these operations, when the aircraft skidded on landing, luckily without casualties. In March 1953 the first shore station, the Royal Naval Air Station at Lossiemouth received its Dragonfly Flight.

The Fleet Air Arm's faith in the reliability of the Dragonfly over earlier helicopters was adequately demonstrated when the aircraft carrier H.M.S. CAMPANIA sailed in June 1952 for the Monte Bello Atomic tests. Two Dragonflies were tasked into the area of the exploision, the first an hour after the burst to collect water samples using a container on a piece of rope. The second to fly over ground zero to collect rocket heads which had been fired through the mushroom cloud. Both missions were completed without a hitch.

The Dragonfly pioneered the early trials at sea proving that air sea rescue by helicopter in atrocious weather conditions was feasible. It proved that operation at sea from fixed wing carriers and from small ships was operationally possible. After 17 years of valuable service and pioneering it was honourably retired from Naval service as it came in, with little fuss or bother.

A dramatic photograph of a Dragonfly Mk 3 accident off HMS GLORY.

Chapter 3

Soldiers and Submarines

Whilst the Dragonfly was busily training pilots and carrying out trials which were proving the helicopters future, the next generation of Fleet Air Arm helicopter was being prepared in the wings — the S-55, more commonly known as the 'Whirlwind'. The Royal Navy's intention was in fact to call this aircraft the CRANEFLY. Happily this name was never used and the name WHIRLWIND was adopted direct from the Americans.

As already mentioned the prototype Whirlwind (S-55) first flew on 10 November 1949. All the American Armed Services were interested in this helicopter which had a far better performance than all existing types. Many different versions were to be produced for the American market and the Admiralty after a lot of deliberation and persuasion of the Government eventually received twelve HRS-2s (Royal Navy Designation HAR Mk 21), and fifteen HO4S-3s, (Royal Navy designation HAS Mk 22), under the Marshall Development Aid Plan (M.D.A.P.).

The HAR Mk 21's were received during 1952 and the first Royal Navy Front-line helicopter squadron, 848 Squadron, was formed on 29 October 1952 with ten HAR Mk 21s. The Malayan

Whirlwind HAS Mk 7 of 845 Squadron alongside in Malta onboard HMS Perseus.

Emergency was getting into full swing and the Government decided to send 848 Squadron to Malaya with all speed. Accordingly the Squadron embarked in H.M.S. PERSEUS and sailed on 9 December 1952. Disembarkation, after Christmas at sea, was on 8 January 1953 to Royal Naval Air Station Sembawang on Singapore Island.

The Work-Up, to becoming operational, began on 12 January with the aim of an operational date of 2 February. However, due to all the R.A.F. S-51's being grounded for mechanical reasons, this date was brought forward to 24 January. A remarkably short work-up period considering the task before them and the experience level within the Squadron. On 24 January 1953, 848 Squadron was deemed operational with three aircraft in place at Kuala Lumpur, three aircraft at R.N.A.S. Sembawang plus four Front Line in Reserve (F.I.R.) aircraft and the majority of maintenance personnel also at Sembawang.

From 24 January two aircraft were at one hours notice from dawn to dusk at both Kuala Lumpur and Sembawang. The first operational task for the Kuala Lumpur detachment was a casualty evacuation of 3 soldiers from a clearing in the jungle to British Military Hospital Kiurara on 25 January. The Sembawang detachment flew their first operational sortie on 27 January, again a casualty evacuation (Casevac), winching out an injured Gurkha soldier and taking him to the British Military Hospital in Singapore.

During 1953 848 Squadron pioneered supply and personnel dropping by parachute from the Whirlwind plus Commando Squadron night flying operations. The first two large scale operations in which the Squadron took part were Operations BAHADUR and CATO. In the first operation 75 Gurkhas were lifted into a Landing Zone using five aircraft (20 sorties). In the second, the first brigade scale operation, 848 Squadron positioned more than 650 soldiers and 4000 lbs of equipment, (183 sorties using 103 aircraft hours), in an area around Benta in Western Pahang in response to an intelligence tip-off that a gathering of Communist delegates was expected in the area. This second operation was not entirely successful from the ground forces point of view as few Communist Terrorists were killed or captured. But it had demonstrated the capability of a few helicopters to rapidly deploy a considerable number of troops. 848 Squadron had arrived.

Back at R.N.A.S. Gosport the HAS Mk 22s were beginning to arrive and were formed into a unit, 706 Squadron, attached to 705 Squadron. These aircraft were equipped with an American

'Dunking Sonar', the AN/AQS4, a short range active sonar. Sonar, equipment for detecting submarines, was no stranger to the Royal Navy as it had been in use in surface ships for many years. The advantages of placing a sonar set in a helicopter were numerous, the most important being that of a stable platform for the equipment. When operated in surface ships sonar suffered from ambient and self-induced noise at high speeds and in rough sea states. (Appendix B contains a description of how sonar works and its application in helicopters).

706 Squadron was commissioned on 7 September 1953 and promptly detached from R.N.A.S. Lee-on-Solent to R.N.A.S. Eglinton in Northern Ireland. Opportunities to exercise with submarines were far more numerous at Eglinton, due to the better weather factor, deep water areas, and the proximity of the Clyde submarine base. This first trial period, including an embarked period on board H.M.S. PERSEUS, was extremely successful and the decision was taken to form the first front line Anti-Submarine Warfare Squadron, 845 Squadron. On 1 March 1954 845 Squadron commissioned at R.N.A.S.Eglinton equipped with the 8 Whirlwind Mk 22s from 706 Squadron. 706 Squadron was disbanded, not to re-emerge again until 1962.

One of the intended uses for the Whirlwind Mk 21 had been to replace the Dragonflies on the carriers as the Search and Rescue aircraft as the Mk 21 had a greater range and payload. Unfortunately the continuation of the Malayan Emergency prevented this happening and the Dragonfly was retained as the seaborne S.A.R. aircraft for longer than anticipated. However the Royal Navy did have one Mk 21 spare which was allocated to Minesweeping trials. Trials carried out in the United States indicated that minesweeping by helicopter was feasible. The Mk 21 was tasked and equipped and the trials took place from RAF Martlesham on the East Coast. Results were not all that promising and after the 1954 trials the project was dropped, but not forgotten.

Now that helicopters using 'dunking sonar' were a proven success the Royal Navy cast around for a helicopter which could overcome a problem area which had emerged. A Whirlwind HAS Mk 22 could only carry the dunking sonar equipment, it could not carry a weapon, in this case — a torpedo, with which to destroy the submarine. This was a severe disadvantage as it meant that a ship had to be called in to attack any sonar contacts gained. The problem was given to Westland Aircraft Ltd and the solution eventually emerged in

1957 in the shape of the Whirlwind HAS Mk 7. This aircraft and other Whirlwind derivatives are examined more closely in Chapter 5.

By 1954 the potential of the helicopter was being fully exploited by the Fleet Air Arm. 848 Squadron was operating in Malaya with Whirlwind HAR Mk 21s and 845 Squadron was in Malta with its Whirlwind HAS Mk 22s continuing to build up experience in A.S.W. tactics. Dragonfly Search and Rescue Flights were operating successfully from both carriers at sea and from shore bases. H.M.S. PROTECTOR the support ship for the British Antartic Expedition was being modified to take a Flight of two Whirlwind H.A.R. Mk 1s. 705 Squadron by 1954 had expanded its types of training aircraft; Hiller H.T.E. 2's had started to arrive in the summer of 1953, the complete order being for 20 aircraft. First flown in 1948, the Hiller H.T.E. 2 had no equal as a training aircraft. It was easy to maintain, and cheap as it was in virtual mass production for the American Armed Services. The Hiller H.T.E. 2 enabled students to easily grasp basic techniques before moving on to the more complicated operational aircraft, the Dragonfly and Whirlwind.

The Fleet Air Arm helicopter force was to remain virtually stationary from 1954 to 1957 whilst successive governments cut the Services down to what they envisaged was the correct size after the Korean War. Armed conflict seemed to have ended for the time being.

Just prior to and during this period some development aircraft were considered by the Royal Navy for service in the A.S.W. role. Although they never entered service with the Fleet Air Arm they are sufficiently interesting to be the subject of the next chapter.

CHAPTER 4

Interesting Prototypes

In 1948 the United States Navy had announced a competition for the next generation of A.S.W. helicopters. A major requirement was that the helicopter should be able to carry the dunking sonar and weapons, thus making it an autonomous unit. In June 1950 the winner was announced — the tandem rotor Bell Model 61. Three evaluation aircraft were ordered by the U.S. Navy and the Admiralty watched the progress of the aircraft with mounting interest.

However the Bell Model 61 was not the only development that was being considered by the Admiralty. In 1948, the Ministry of Supply had ordered two prototypes of a tandem rotor helicopter (designed to a Ministry of Supply specification E4/47), to be built by the Bristol Aeroplane Company.

The Bristol aircraft, Type 173, first flew on 3 January 1952. It was powered by two 550 h.p. Alvis Leonides engines either of

The unsuccessful Bristol 173 carrying out trials onboard HMS EAGLE.

which was capable of taking over and providing power to the other rotor in the event of a single engine failure. It was only capable of maintaining straight and level flight at low All Up Weights with one engine out. This performance was to be improved by the installation of Bristol Janus turbine engines, but in fact they were never installed. With the promise of better performances later, the Royal Navy maintained its interest and in 1953 carried out trials with the first prototype in H.M.S. EAGLE. These trials were adjudged a success but certain modifications to the aircraft were required before it could be accepted for Naval Service. The undercarriage had to be modified to allow the carriage of torpedoes, and as it was too long to go down the standard aircraft carrier lift its airframe had to be shortened. This modified version known as the Mk 5 was built but never flown.

Meanwhile the Bell Model 61 (now designated HSL-1) had made its first flight on 4 March 1953. Its 4000 lbs of payload made it an extremely attractive proposition especially as it had been specifically designed as a carrier-borne A.S.W. helicopter. Dunking sonar, torpedoes and depth charges were an integral part of the package. In 1953 the Admiralty ordered 18 HSL-1's, the U.S. Navy having ordered 60. The Fleet Air Arm was now in a good position with 18 of one advanced type of helicopter ordered and promising trials being conducted on a similar British type. Deliveries of the HSL-1 were delayed until 1956 as the test programme was taking longer than expected. This was mainly due to the development and installation of an auto-pilot system developed so that the aircraft could operate both by day and by night, (All-Weather operation).

By the end of 1954 the Korean War had ended and the Government, in order to save dollars, cancelled the HSL-1 order.

However the Fleet Air Arm still had the on-going Bristol 173 programme which appeared to be progressing satisfactorily, more trials having been carried out at sea in 1954.

By the end of 1954 a production order for 65 Bristol helicopters was placed, the version was to be called the Type 191. The intention was to have the Type 191 in Squadron service by 1957.

Unfortunately by the end of 1956 it was becoming obvious that the Type 173/191 had serious engine problems and in the 1957 Defence White Paper the Royal Navy order was cancelled. The Fleet Air Arm was left with no new aircraft which could carry both the dunking sonar and weapon. However, as mentioned earlier, Westlands had been asked to study the problem and to investigate modifying the Whirlwind to carry at least one or the other so the prospects were not entirely grim.

1957 saw the end of a period during which the Fleet Air Arm had had high hopes of receiving at least one very advanced A.S.W. aircraft. Due to lack of money, poor initial design and Defence cuts the Fleet Air Arm found itself no better off than it had been in 1954 when the first Front Line Whirlwind squadron had been formed. Three years development had been lost and those years took a long time to recover.

CHAPTER 5

The Ubiquitous Whirlwind

In 1953 Westland Aircraft Ltd had obtained a licence to build and modify as necessary the S-55 Whirlwind. The first series of aircraft produced by Westland, the Series I, were Westland built airframes powered by American engines. The H.A.R. Mk 1 was powered by a 600 h.p. Pratt and Whitney R-1340-40 Wasp. In 1955 the Royal Navy received ten Mk 1s and allocated seven to 848 Squadron to replace the by now ageing Mk 21s which were withdrawn and converted to the A.S.W.

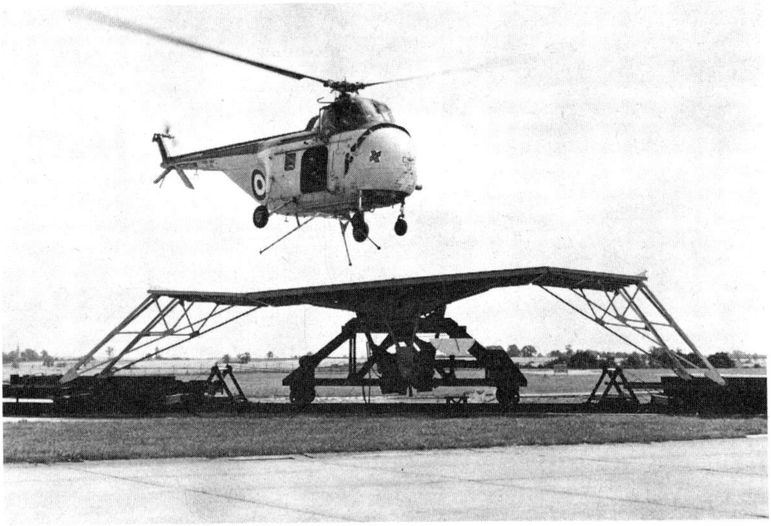

Whirlwind HAR Mk 1 being used to establish rates of descent on the Rolling Platform at RAE BEDFORD.

role. Two Mk 1s went to H.M.S. PROTECTOR flight, as already mentioned, and the last aircraft was maintained in reserve. These were the only Mk 1s to serve in the Fleet Air Arm and gave good, if under-powered, service.

The second aircraft in the Series 1 was the HAR Mk 3. This aircraft had the more powerful 700 h.p. Wright R-1300-3 Wasp engine and 25 were bought, coming into service also in 1955. The Mk 3s would at last replace the Dragonflies in the Search and Rescue role ashore and afloat. The first carrier flight to receive Mk 3s was that of H.M.S. ARK ROYAL in Summer 1955. Other flights converted as aircraft and crews became available.

1956 saw Westlands working urgently on the already

Whirlwind HAS 22 of 845 Squadron embarked onboard a Light Fleet Carrier.

mentioned problem of the torpedo carrying Whirlwind. The tactic of waiting for a ship or a weapon carrying Gannet to come and assist in any A.S.W. action plus the short endurance of the A.S.W. Whirlwind *and* the improved performance of submarines was beginning to prove very embarrassing. The answer at last flew on 17 October 1956, the Whirlwind H.A.S. Mk 7.

The Whirlwind HAS Mk 7 was a development of the HAR Mk 5, an aircraft designed to eventually replace the Mk 1 and Mk 3. Only two Mk 5s were built and neither entered naval service.

Whirlwind HAR Mk 3 of 701 Squadron at RNAS Lee on Solent modified for Minesweeping Trials. Note the motor-cycle type silencer.

The HAS Mk 7 was a Series 3 Whirlwind and was powered by a 750 h.p. (derated) Alvis Leonides Major Mk 5 engine. It had a drooped tail to give greater clearance between main and tail rotor blades when landing under difficult conditions such as on a small deck or under auto rotation conditions. The Mk 7 only differed from the original Mk 5 design in having a raised cabin floor and the under floor fuel tanks repositioned. These differences gave the Royal Navy what it had been asking for — a torpedo bay underneath the aircraft.

On 18 March 1957 700H Squadron, the Intensive Flying Trials Unit (I.F.T.U.) formed at R.N.A.S. Lee-on-Solent to prove and introduce the HAS Mk 7 into service.

Trials were successful and in June 1957 845 Squadron discarded its eight Mk 22s and re-equipped with eight HAS Mk 7s. One hundred and thirty HAS Mk 7s were ordered by the Royal Navy.

Now at last the Fleet Air Arm had a helicopter to carry the sonar or the weapon but not unfortunately both. The system and aircraft were proven and it was decided to replace the A.S.W. Gannet Squadrons by HAS Mk 7 Squadrons. During

Whirlwind HAS Mk 7 of 825 Squadron onboard HMS VICTORIOUS.

1958 three Squadrons re-equipped at R.N.A.S. Eglinton. 820 Squadron in January, 824 Squadron in April and finally 815 Squadron in September.

It is worthwhile to digress slightly at this point from the Whirlwind story and take a look at a short lived but extremely interesting and busy Squadron which formed at R.N.A.S. Lee-on-Solent on 31 October 1957, 701 Squadron. 701 Squadron was tasked with being responsible for the Work-up and Continuation Training of all Ships Flights on their formation and during their disembarked periods. It also provided Trials, Fleet Requirements, Communications and Search and Rescue aircraft for the Portsmouth area. Most of these tasks were taken over from 705 Squadron. Headquarters Flight consisted of two Dragonfly Mk 5s, one Whirlwind Mk 3, one Whirlwind Mk 22 (V.I.P.) and one Whirlwind HAS Mk 7. Four flights were established, A Flight (H.M.S. EAGLE), B Flight (H.M.S. ARK ROYAL), C Flight (H.M.S. ALBION) and D Flight (H.M.S. BULWARK). All flights were by this time equipped or re-equipping with two Whirlwind HAR Mk 3s. It was intended that the Hiller H.T.E.-2 in H.M.S. VIDAL (to be later replaced by a Dragonfly) and the two Whirlwind HAR Mk 1s in H.M.S. PROTECTOR would eventually be absorbed into the Squadron. During its short life, until 23 September 1958, when it was axed by that years Defence White Paper, it carried out many and varied tasks. Minesweeping Trials were conducted with a Whirlwind HAR Mk 3, again with no conclusive results. The minesweeper H.M.S. BURNASTON was towed successfully using the helicopter sweep gear. On 6 August 1958 a Whirlwind HAR Mk 3 towed a Slingsby Glider aloft as a rehearsal for an Air Day. This was the first time this had been done in Britain, but not in Europe, as it was reputedly achieved in Poland in the summer of 1957. Sonar Type 194 trials were carried out in the summer of 1958. This Dunking Sonar was an 'anglicised' version of the American AN/AQS 4 already in use in 845 Squadron. 'Black boxes', that is the electronic control boxes, were made interchangeable between the Type 194 and AN/AQS 4 which was quite cost effective for the Fleet Air Arm. During its short history 701 Squadron also took part in many Search and Rescue operations from looking for escaped criminals to lifting injured seamen from merchant and naval ships. Newspaper cuttings from local newspapers indicate that the civilian populations from the nearby towns of Fareham, Gosport and Lee-on-Solent really took 701 Squadron into their hearts during its short life span.

R.N.A.S. Lee-on-Solent was becoming crowded what with both fixed and rotary wing aircraft based there and with the increased demand for helicopter pilots and aircrewmen more and more flying was necessary. Therefore in January 1960 705 Squadron moved down to R.N.A.S. Culdrose in Cornwall, where it has been based ever since. At the time of the move it was equipped with Dragonflies, Whirlwinds of all marks and the Hiller H.T.E.-2. The Dragonflies did not have long to go though, and the last four were phased out of 705 Squadron in March 1962. Although R.N.A.S. Culdrose was also a fixed wing air station the surrounding area was relatively free of airspace restrictions and the satellite airfield of Predannack gave 705 Squadron the use of an additional landing area not available at Lee-on-Solent.

An important event in 1959 was the start of the conversion of H.M.S. BULWARK, formerly a fixed wing carrier, to the Commando Carrier role. The 1958 Defence White Paper had spelt out the Government policy of creating an amphibious force to combat 'brushfire wars' and the Bulwark conversion was the first sign of that policy.

The concept of the Commando Carrier had been amply demonstrated during Operation MUSKETEER, the Anglo/French landings at Suez in 1956. 845 Squadron's ten Whirlwind HAS Mk 22 were converted into the troop carrying role and embarked in H.M.S. THESEUS. The Joint Experimental Helicopter Development Unit, a joint R.A.F./Army unit with Whirlwinds and Sycamores, was embarked in H.M.S. OCEAN. (The word Experimental was dropped during the Operation). At dawn on 6 November 1956 the world's first large scale helicopter invasion started with 845 Squadron and the Joint Helicopter Development Unit taking ashore 45 Commando, Royal Marines. The landing was a success, but it must be borne in mind that a fixed wing and ship bombardment had been going on for a few days previously. Also a conventional seaborne launching took place an hour before the first 'wave' of helicopters. The first Marine casualty was being operated on less than one hour after he had eaten breakfast aboard the same carrier, a fact demonstrating yet another role for the helicopter, that of casualty evacuation. One of 845 Squadron's aircraft was lost after running out of fuel and several aircraft were hit by small arms fire. Lessons had been learnt from this operation and the result, as already mentioned, was in the 1958 Defence White Paper.

With the dramatic increase in the numbers of A.S.W.

helicopter Squadrons, and their requirement to operate with submarines and ships, operations from R.N.A.S. Eglinton in Northern Ireland, were becoming less viable, (mainly due to lack of ships). The Naval Base at Portland near Weymouth in Dorset, where it had been decided to start a Work-Up base for ships, appeared to also suit the purposes of an A.S.W. helicopter base. There had been a helicopter pad there since the days of the R-4B, but the intention now was to build a purpose-built helicopter station, the first in Europe.

On 24 April 1959 R.N.A.S. Portland was commissioned and its first resident squadron was 815 Squadron which moved down from R.N.A.S. Eglinton. The Whirlwind HAS Mk 7 had been going through a traumatic period since its introduction in 1957 due to a spate of engine problems. During the summer of 1959 all Whirlwind HAS Mk 7s were grounded, apart from essential flights, in order that modifications could be carried out. 815 Squadron ditched one of their aircraft three days before the Portland commissioning and another five days after the commissioning. Luckily few ditchings were fatal as most engine failures occurred when the engine was at full power in the hover and normally the crew was able to get out safely.

Up until the middle of 1959 the training for helicopter Anti-submarine pilots, observers and sonar operators was largely carried out after they had joined their Squadron. Basic flying training taught the basic skills of flying and navigation but no tactical thinking or equipment operating. With the ever increasing complexity and variety of tactics and equipment it was thought necessary to set up an Operational Flying School. On 28 August 1959 737 Squadron was commissioned at R.N.A.S. Portland with Whirlwind HAS Mk 7s and two HAR Mk 3s for Search and Rescue. By May 1960 737 Squadron had built up to full strength of nine HAS Mk 7s and two Mk 22s, these last being more reliable had replaced the Mk 3s. Students joined R.N.A.S. Portland for courses at the Naval Air Anti-Submarine School using the 737 Squadron aircraft for the practical flying exercises.

In 1960 two further Whirlwind HAS Mk 7 Squadrons commissioned, 814 Squadron in May and 825 Squadron in August. Both Squadrons replaced Gannet squadrons at R.N.A.S. Culdrose. In December of the same year 815 Squadron, having finished H.M.S. ALBION'S last commission prior to her conversion to a Commando carrier, disbanded. 820 Squadron had already disbanded in October that year having no carrier to go to.

None of the Whirlwind HAS Mk 7 Squadrons lasted a long time. 814 Squadron disbanded in September 1961, 825 Squadron in April 1962, 737 Squadron Mk 7s went to 771 Squadron in July 1962 and the last operational Mk 7 anti-submarine squadron, 824 Squadron, disbanded in May 1963. The majority of Squadrons converted to the Wessex HAS Mk 1 which is fully covered in the next chapter.

The Whirlwind Mk 7 did continue in operational service in the Commando role in 846 Squadron, from November 1962 to October 1964, in 847 Squadron from May 1963 to December 1964, and with 848 Squadron until July 1963 when the squadron disbanded. 846 Squadron embarked in the new Commando Carrier, H.M.S. ALBION, and went out to the Far East with 845 Squadron to participate in quelling the Brunei revolt and in the beginning of the Indonesian Confrontation in 1963. 847 Squadron remained at R.N.A.S. Culdrose as the Headquarters and Training Squadron for Commando Squadron pilots.

Whirlwind HAS Mk 7s remained in the Fleet serving in 771 Squadron from August 1962 to December 1964 in a Trials, Search and Rescue and Fleet Requirements role. On the disbanding of 771 Squadron in December 1964 the Mk 7's were transferred to 829 Squadron to continue performing the same tasks. In June 1967 771 Squadron was reformed at RNAS Portland and the Mk 7s were transferred yet again. Here they remained until 1970 when they were replaced finally by Wessex HAS Mk 1 aircraft. The last few Mk 7s continued on in 705 Squadron at R.N.A.S. Culdrose in the training role until 1975 when they were replaced by the Gazelle and at last the Whirlwind HAS Mk 7 went into retirement — but not quite.

Between 1959 and 1960 Westland Aircraft Ltd had converted four early Whirlwinds to take the 1024 s.h.p. de Havilland Gnome turbine engine giving an impressive increase in performance and reliability. The Royal Navy took an interest in this improved Whirlwind called the HAR Mk 9, and as a number of Mk 7s were becoming surplus to requirements it decided initially to convert seventeen Mk 7s to HAR Mk 9 standard. The object was to use the new aircraft in the Search and Rescue role plus two HAR Mk 9s would replace the Whirlwind Mk 1s in H.M.S. PROTECTOR.

Conversion of the Mk 7s started in September 1964. H.M.S. HERMES Search and Rescue Flight were the first unit to receive Mk 9s in the summer of 1966. H.M.S. PROTECTOR's Flight was next followed by the Search and Rescue Flights at RNAS Culdrose, Brawdy, Lossiemouth and Lee-on-Solent.

A Whirlwind Mk 9 from HMS ENDURANCE operating in the Antartica in support of the British Antarctic Survey.

These converted Mk 7s continued on (the two in H.M.S. ENDURANCE, which replaced H.M.S. PROTECTOR, being replaced by Wasp HAS Mk 1 in September 1976), until April 1977 when the last Mk 9 flew from R.N.A.S. Lee-on-Solent. By this time the air stations at Lossiemouth and Brawdy had been handed over to the RAF. Search and Rescue was carried out from RNAS Culdrose by 771 Squadron Wessex Mk 1s and at RNAS Lee-on-Solent a Wessex HU-5 replaced the Mk 9.

The Whirlwind in seven different disguises served the Fleet Air Arm from 1952 until 1977, a total of 25 years. It was never ideal for any of the many different tasks it undertook, but it was versatile as were the aircrew who flew it. It did build up the expertise that the Fleet Air Arm required during the 1950's and early 1960's and it prepared the way for the next generation of 'electronic' aircraft that were to follow. The Whirlwind was a very adequate stepping stone.

An early Whirlwind Mk 3 accident at Gosport Hampshire. The engine failed, at night, and during the landing the tail was chopped off.

CHAPTER 6

The Night Fliers

When the 1957 Defence White Paper cancelled the Bristol Type 173/191 order the Royal Navy was left with no option but to convert the Whirlwind to give the Fleet Air Arm a limited search and strike capability with the HAS Mk 7.

The United States Navy meanwhile, having not pressed on with the Bell HSL-1, had turned its attention to a new Sikorsky helicopter, the S-58. This aircraft had first flown on 8 March 1954 and entered U.S. Navy service in late 1955 as the HSS-1. Further developments in the United States introduced auto-stabilisation (auto-pilot), day and night instrumentation, better navigation aids and the most important improvement of all, the hover coupler. The hover coupler was designed to take an aircraft from a 'gate' position down to a pre-selected height above the water and then hold the aircraft in a motionless hover whilst the dunking sonar body was lowered into the water. An accurate manually controlled hover over the dunking sonar body is not possible at night, due to the lack of external references, and the hover coupler and associated equipment

An American HSS-1 bought by Westlands for pre Wessex Mk 1 trials.

was designed to do this electronically. By day the system relieved the pilot(s) of the strain of a manually controlled hover. The technology required to achieve a night hover was very advanced and therefore in the early days the system was relatively unreliable and had to be constantly monitored.

The Royal Navy, having lost the Bell HSL-1 and the Bristol Type 173/191, considered the Sikorsky HSS-1 the only feasible helicopter available at the time to replace the ageing Whirlwinds, whose operational capabilities were fast being outstripped by improved submarine performances. Also the Commando Whirlwind version was becoming obsolete in terms of performance.

In 1956 Westland Aircraft Ltd obtained a licence to build the Sikorsky S-58 in Britain but not with the original Wright R-1820-84 engine. Turbine engines were now at the size and reliability required for over-sea operations. One S-58 was brought from America and modified to take a 1,100 s.h.p. Napier Gazelle turbine engine. It successfully flew for the first time on 17 May 1957. The Royal Navy was extremely pleased with this new aircraft and two pre-production aircraft were built for Naval Trials, the first flying on 20 June 1958. This new aircraft was called the Wessex HAS Mk 1.

Full scale production started in 1959 and in June 1960 the Intensive Flying Trials Unit, 700(H) Squadron, formed at RNAS Culdrose. The object of the I.F.T.U. was to prove the aircraft, especially its night flying equipment. The sonar was the same as that carried in the Whirlwind HAS Mk 7, with proven reliability and performance.

The Wessex HAS Mk 1 differed in many respects from all preceding Fleet Air Arm helicopters. The main differences being its turbine engine, the auto-stabilisation equipment (A.S.E.) and a night dunking capability. It also had a vastly improved performance in speed, endurance and very important — payload.

Before progressing into a more detailed description of the Wessex HAS Mk 1 equipment and its method of operation it is necessary to look at the only piece of equipment that was

814 Squadron Wessex HAS Mk 1 helicopters taking off for an ASW screening exercise.

common to both the Wessex HAS Mk 1 and the Whirlwind HAS Mk 7. This particular piece of equipment enabled both Whirlwind and Wessex to maintain an accurate hover over the dunking sonar body whilst it was in the water. The reason for requiring an accurate hover is that the sonar 'beam', (like that emitted from a torch), required the body to be vertical and stationary as far as possible. If the cable was dragged at all it could mean that the beam would be actually pointing upwards or downwards, seriously degrading the efficiency of the sonar search. In the Whirlwind HAS Mk 7 the sonar cable after rolling off the sonar winch drum passed through two 'forks' at right angles to one another. Provided the cable was accurately up and down i.e. vertical, the pilots indication, a meter with two needles at right angles to each other, representing a plan position of the two 'forks', showed the two needles crossing in the centre of the meter. Any movement of the cable away from the vertical would be indicated by the needles moving away from the centre in the same direction as the cable movement. The aircraft height above the water was calculated electrically by taking the depth of the sonar body, (from a pressure sensing unit), from the length of cable paid out. All corrections in the Whirlwind HAS MK 7 were made manually by the pilot flying the aircraft. In the late 1950's/early 1960's operating with slow speed submarines in good water conditions could force a Whirlwind HAS Mk 7 to remain in the hover for periods of up to an hour. This gave the pilot considerable problems as he was continually altering the aircraft power requirements using a twist grip throttle in his left hand and getting cramp in his left leg from having to apply a constant pressure on the rudder pedal to keep the aircraft heading into wind. It was very seldom the Whirlwind HAS Mk 7 flew with two pilots due to payload and fuel considerations. The Wessex achieved a stable hover electronically but used the same basic equipment.

The helicopter has to face into the wind whilst entering and leaving the hover. The resultant 'S' shaped flight profile from hover to hover did not lend itself to any particular navigational technique, and navigation during anti-submarine operations was very much dependent upon the experience and intelligence of the Observer. Whirlwind A.S.W. Squadrons did very well to achieve the results that they did. The Wessex HAS Mk 1 was to improve this facet of operation as well.

Now to look at the Wessex HAS Mk 1 in detail; the Fleet Air Arm's first 'electronic' helicopter. Firstly the Auto Stabilisation Equipment (A.S.E) and the Flight Control System (F.C.S). The A.S.E. gave the pilot(s) the ability to fly the aircraft 'hands off'. With the aid of various 'trims' he could trim the aircraft in level flight and let the A.S.E. do all the work. In practice one pilot always had his hands 'hovering' over the controls in case of a malfunction. Height control was achieved by fitting a Barometric Hold which could maintain the aircraft at the pressure altitude at which the hold had been engaged. A Radio

Altimeter Hold which operated between sea level and five hundred feet above sea level was also fitted. The Radio Altimeter was a necessity as the anti-submarine mode transit height of the Wessex HAS Mk 1 was only 125 feet and the Barometric Hold was capable of large fluctuations in height.

During daytime, in order to speed up the time getting from one hover to another the pilot flew the aircraft manually into the hover. At night, when the aircraft always had two pilots, the F.C.S. (Flight Control System) came into its own. The aircraft would be flown to a 'gate' which had to be satisfied by several conditions — facing into the wind, at operating height (125 feet) and at an airspeed of 90 knots. The flying pilot would already have fed into the F.C.S. the hover height he required, dependent upon the sea state (40 feet was normal) and once the 'gate' conditions were satisfied he would order the non-flying pilot to 'Engage Transition Down'. Once the non-flying pilot had pressed the requisite button the F.C.S. computer would automatically bring the aircraft to the hover at the preset lower height at zero ground speed. Height indication would be fed into the computor from the Radio Altimeter and speed indication from two Doppler aerials, one facing forward and one aft underneath the aircraft. The computer would make all control and throttle movements based on the Radio Altimeter and Doppler inputs during the Transition Down. Once established in the hover the sonar body was lowered into the water. Almost immediately the non-flying pilot would switch from the part of the F.C.S., still being run by height and speed indications, to another part of the system called the Cable Mode. This involved taking inputs from the horizontal forks through which the cable passed (as in the Whirlwind) and electronically, through the F.C.S., maintaining the forks in a central position. Hover height was maintained also by the F.C.S. using the depth pressure sensor and amount of cable out indications (as in the Whirlwind). Leaving the hover was exactly the reverse of the above sequence.

The Auto Stabilisation Equipment and the Flight Control System were produced by Louis Newmark and were their first attempt at producing such a system. When the systems were working well the results were very good. However it was a Simplex system, i.e. no duplication of wires or electronic control boxes, and a failure in one or more components normally meant a cancelled flight or at worse a swim for the aircrew. Many day sorties were flown simulating component

A Wessex Mk 1 undercarriage being repaired at Sibu airfield in Borneo. The strut failed after a landing near the Indonesian border. The repair was completed in eight minutes.

failures in order that pilots would feel and see the effects and be able to react correctly to any failures, especially at night.

Any gain from the improvements described above would have been nullified by inaccurate 'jump navigation' as practised in the Whirlwind HAS Mk 7. Therefore the Intensive Flying Trials Unit was tasked with finding an accurate system of navigating from hover to hover i.e. Jump Navigation. Firstly this required standard flight profiles; angles of bank, airspeed and acceleration/deceleration had to be standard. Three triangulation points were set up overlooking Falmouth Bay near R.N.A.S. Culdrose, and for days on end the I.F.T.U. Wessex HAS Mk 1 'jumped' around the Bay noting all relevant detail. The result was a set of graph tables based on windspeed, the set was divided into two. One half for Manual Jumping by daytime, i.e. without the assistance of the Flight Control System, the other for night time jumping. Provided the pilot flew the standard flight profile and the correct graph had been selected the accuracy in Jump Navigation was very high.

700(H) Squadron continued its trials from June 1960 until July 1961 when it was disbanded and its aircraft passed on to the first Wessex HAS Mk 1 Squadron, 815 Squadron, which commissioned at R.N.A.S. Culdrose on 4 July 1961. Westland Aircraft Limited had been producing the Wessex HAS Mk 1 at Yeovil since late 1959, and the re-equipping of other Squadrons and the formation of new Squadrons went ahead swiftly. 819 Squadron formed at R.N.A.S. Eglinton with 2 Wessex HAS Mk 1 and 1 Whirlwind HAS Mk 7 on 5 October 1961 and 814 Squadron re-equipped with 6 Wessex HAS Mk 1 on 29 November 1961.

The helicopter fleet of the Fleet Air Arm was now starting to expand especially on the aircrew side. The introduction of the Wessex HAS Mk 1 had doubled the pilots required, and it had become obvious that the system of training pilots in 705 Squadron on Whirlwind HAS Mk 7, a piston-engined helicopter, and then sending them straight to their Front Line Wessex HAS Mk 1 Squadron would not work.

Accordingly, in early 1962 706 Squadron was reformed at R.N.A.S. Culdrose with six Wessex HAS Mk 1 to become the Advanced Flying Training Squadron for Wessex pilots. These Wessex pilots were not only destined for Anti-Submarine Squadrons. The advantages of using the Wessex in the Commando role had been fully realised, (greater payload, greater range and endurance). 845 Squadron was reformed, (having disbanded in 1959 when H.M.S. BULWARK commenced her conversion to a Commando Carrier), on 10 April 1962 as a Commando Squadron with 12 Wessex Mk 1. The Wessex Mk 1 supplied to the Commando Squadrons were the same as the A.S.W. Wessex except for the removal of the sonar and the Flight Control System. The Auto Stabilisation Equipment was retained. The aircraft kept the capability to fit sonar and both H.M.S. BULWARK and H.M.S. ALBION did carry the Type 195 Sonar for fitting if required.

The Wessex HAS Mk 1 continued as the Fleet Air Arm's major A.S.W. helicopter from 1961 until 1967. In the Commando role the Mk 1 was superceded some three years earlier, in the autumn of 1964, by the Wessex HU Mk 5.

The Wessex HU Mk 5 was derived directly from the Wessex Mk 1, the major difference being the replacement of the single Gazelle Mk 161 engine by two coupled Gnome engines. The airframe and rotor head of the Mk 5 were strengthened to cope with the extra power produced. The result was an extremely versatile Commando support aircraft with a much improved lift

A Wessex HU 5 of 707 Squadron rocketting at Castlemartin Range in South Wales.

capable of carrying underslung Land Rovers to a distance of 25 miles and returning to base without refuelling. Its twin engines gave it greater reliability, and that plus a stronger airframe and rotor head gave the aircraft a potential for more severe manoeuvres at low level where the likelihood of enemy small-arms fire was high. The Wessex Mk 1 in the Commando role had three retaliatory weapons at its disposal, 2 inch rockets carried on pods outside of the undercarriage struts, the Fixed Forward Firing Gun (F.F.F.G.) carried in a similar position, and the SS11, a wire guided anti-tank missile.

The HU Mk 5 increased this inventory with the fitting of the A.S.12, a potent wire guided anti-ship missile. All marks of Wessex were capable of fitting the General Purpose Machine Gun (G.P.M.G.) in the cabin door and window. This weapon was very limited in its application as in order to fire it the aircraft had to maintain a side-on attitude to the target area thereby presenting itself as an ideal target.

The Wessex HU Mk 5 Intensive Flying Trials Unit, 700(V) Squadron, formed at R.N.A.S. Culdrose on 5 December 1963 with six Wessex HU Mk 5 aircraft. After an extremely successful trials period it disbanded in October 1964. 848 Squadron re-equipped with 18 Wessex HU Mk 5 that same October, 845 Squadron following suit in October 1965 with 18 aircraft.

The introduction of the Wessex HU Mk 5 with a different engine system and more sophisticated flying techniques led to the formation of a Commando Advanced Flying Training Unit, 707 Squadron, at R.N.A.S. Culdrose on 9 December 1964. By May 1966 707 Squadron had built up to 13 aircraft.

Two further Commando Squadrons formed with Wessex HU Mk 5 aircraft. In Autumn 1968 846 Squadron formed at R.N.A.S. Culdrose as a Headquarters Squadron with five aircraft. The reason for this move was to relieve 707 Squadron, which was a training squadron, from some of the extraneous tasks and trials it was being called upon to meet. 845 and 848 Squadrons were already heavily committed to H.M.S. ALBION and H.M.S. BULWARK respectively. It also meant that a nucleus of Commando expertise was available on call in the United Kingdom. On 14 March 1969 847 formed at R.N.A.S. Sembawang in Singapore taking 8 aircraft from 848 Squadron. 847 Squadron remained in Singapore until 22 May 1971 providing a continuous Commando Squadron presence in the area and occasionally embarking for exercises or detaching up-country into Malaysia.

A Wessex HU 5 moving troops along the North Borneo border. Note the absence of the sand filter under the nose.

A Wessex HU 5 of 848 Squadron in action during the British Withdrawal from Aden.

No mention of the Commando Squadrons would be complete without commenting on the time that they spent in Borneo during the Confrontation with Indonesia from 1963 to 1966. Three Squadrons were involved at one stage or another, 846 Squadron (6 Whirlwind Mk 7) ashore and on board H.M.S. ALBION, 848 Squadron (12-16 Wessex Mk 1) ashore and aboard first H.M.S. ALBION and then H.M.S. BULWARK, and 848 Squadron (18 Wessex HU Mk 5) ashore and on board H.M.S. ALBION. Detachments were based ashore in such exotically named places as Bario, Labuan, Sibu and Nanga Gaat operating under very makeshift conditions. In many places in Borneo the terrain had never been mapped and it was a matter of making your own as you flew. Maintenance ratings worked under what could hardly be called ideal conditions, indeed at Nanga Gaat all accommodation and even a 'pub', the Anchor Inn, was built out of local materials. Yet, there were very few times when a mission was cancelled or refused due to technical unserviceability. A credit to both the maintenance personnel and the aircraft.

The Wessex HU Mk 5 still continues on in the Fleet Air Arm in 845 and 846 Squadrons with 12 aircraft each, and in 707 Squadron with 13 aircraft (846 Squadron having been reformed

Torpedo dropping in Falmouth Bay by a Wessex HU 5 of 845 Squadron.

in October 1978). 772 Squadron the Search and Rescue and Fleet Requirements Squadron at R.N.A.S. Portland which formed in September 1974, originally with 6 Wessex Mk 1 from 771 Squadron, also operates the Wessex HU Mk 5. Two ships flights — those of R.F.A. RESOURCE and R.F.A. REGENT — have one Wessex HU Mk 5 each; these flights are parented by 772 Squadron. The primary task of these R.F.A. Flights is the replenishment of other ships at sea by Vertical Replenishment (Vertrep). Loads are taken from the flight deck of the R.F.A. suspended underneath the aircraft and landed on board the receiving ship. This method will never replace the existing alongside methods of replenishment at sea but is complementary to them. Two Wessex HU Mk 5 aircraft have been converted to carry V.I.P's, they are part of 781 Squadron at R.N.A.S. Lee-on-Solent and are painted white and green. These aircraft, because of their colour scheme, are affectionately known as 'Green Parrots'.

With a change of emphasis in British Defence Policy from world-wide to N.A.T.O., the Commando Squadrons have become increasingly involved in Arctic type operations. The Northern Flank has become one of the Commando Squadrons areas of interest. So that a sufficient number of aircrew and maintainers trained in Arctic operations are always available an annual exercise called CLOCKWORK is run in Northern Norway, a far cry from the jungles of Borneo.

Whilst the Wessex HU Mk 5 was establishing itself the Royal Navy was finding more uses for the Wessex HAS Mk 1. A new class of Destroyer, the County Class, had been designed with a hangar and Flight Deck aft. The Wessex HAS Mk 1, around which the hangar had been designed, first embarked in H.M.S. DEVONSHIRE, the first of the Class. Eight County Class destroyers were built (H.M.S. LONDON, DEVONSHIRE, KENT, ANTRIM, GLAMORGAN, FIFE, NORFOLK, HAMPSHIRE) and all were equipped with a Wessex Flight. All Flights were parented by 829 Squadron at R.N.A.S. Portland and did a separate work-up to the ship, culminating in an inspection to check that the Flight was ready to go to sea in all respects from flying to maintenance. These Flights are part of the standard armament of the County Class destroyer. In June 1970 the responsibility for parenting the Flights passed from 829 Squadron to 737 Squadron as the mix of aircraft (Wessex HAS 1/Wasp) and the size of the parenting responsibility for aircraft and aircrew was inappropriate for 829 Squadron.

A further development of the Wessex HAS Mk 1 was the Wessex HAS Mk 3, discourteously referred to by aircrew and others as 'the Wessex with the hump' or more simply 'the Camel'. However before going on with the Wessex HAS Mk 3 it is necessary to look at equipment advances in the rear cabin of the Wessex.

During the late 1950's and early 1960's the Admiralty Underwater Weapons Establishment at Portland and an electronics firm — Plessey Limited — had been working on a new advanced dunking sonar, the Type 195. Two other firms were working on other improvements, Ekco on a lightweight helicopter radar and Marconi on a navigation system. The object of the total package was to make the A.S.W. helicopter capable of operating independently by day and by night within the limits of its endurance. Westland Aircraft Limited meanwhile, in conjunction with Louis Newmark, worked to improve the aircraft performance and Flight Control System. There had been a lot of teething troubles with the Wessex HAS Mk 1, free power turbines disintegrating, starter fires, amongst others, and the reliability of the Simplex F.C.S. was not as good as it might have been. In addition although the Wessex HAS Mk 1 was capable of flying with a full crew plus sonar plus a weapon, the much improved capabilities of submarines were demanding ever increasing endurance and operating distances from ships and the actual sortie length was yet again becoming operationally impractical. The total improvements package of sonar, radar, navigation system, engine and F.C.S. improvements were to give the Royal Navy a worldwide lead in A.S.W. helicopter technology and tactics. However the concept of a helicopter which could effectively Search and Strike all in one sortie was not fully realised until the introduction of the Sea King HAS Mk 1, still some four years away.

The advanced sonar, the Type 195, was test flown in a converted Wesex HAS Mk 1 known as 'Bootsie'. 'Bootsie' was so called as it was fitted with an enormous flotation bag down

'Bootsie', a Wessex Mk 1 equipped specially for 'dunking sonar' trials. Whirlwind Mk 7's of 737 Squadron can be seen in background.

either side so that it would float in the event of an engine or control failure. The extremely valuable test equipment and sonar set, which at that time was also highly classified, would then be recoverable. In fact this problem of recovering aircraft which had ditched into the sea had worried the Fleet Air Arm for sometime and all types of Wessex were retro-fitted with a flotation system. This system consisted of an inflated rubber bag in the tail plus flotation canisters on either wheel. The flotation cannisters would inflate automatically on entering the sea looking somewhat like two large doughnuts. In low sea states the system was very successful and many aircraft were recovered intact. The object of recovery was not so much the rescue of the aircraft but the ability to establish the cause of the accident. Provided the flotation gear worked correctly it also gave the crew an added safety factor.

The Wessex HAS Mk 3 first flew with all the improved and new equipment during 1966 and on 9 January 1967 the Intensive Flying Trials Unit, 700(H) Squadron, formed at R.N.A.S. Culdrose with five Wessex HAS Mk 3. Externally the aircraft differed considerably from the Wessex HAS Mk 1. Aft of the rotor head the fairing was extended and just aft of that fairing was the radar radome. From the nose of the aircraft protruded an airspeed indicator probe. Internally the aircraft had changed also. The engine was now a 1,600 s.h.p. Gazelle Mk 165, with a compressed air starting system instead of the highly inflammable AVPIN, a liquid highly disliked at sea due to its inflammable properties. The rear cabin was almost totally filled by the Type 195 Sonar and the new radar, so much so that there was little room for the Observer and the Sonar Operator to squeeze through into their flight positions. The pilots had a completely new Flight Control System developed from the Wessex HAS Mk 1 system. The system was Duplex, that is all systems had a back-up unit or circuit which would take over in the event of a system failure. Furthermore, the Observer in the rear cabin had a limited manual control over the flight profile and indeed it was possible for the pilot to 'fly' from one hover to the next without touching the controls other than pressing two buttons. Unfortunately the inclusion of weapons into the payload made the sortie length only just operationally viable.

Wessex HAS Mk 3 of 820 Squadron embarking onboard HMS BLAKE.

The Fleet Air Arm now had an all-weather helicopter which could operate with complete freedom. It had radar, it had good communications, it had very advanced sonar and it could operate meaningfully with surface units without expecting radar control or navigational information. The limitations of weapons and endurance still remained to a degree but these were being considered and the Naval Staff had already chosen the Wessex HAS Mk 3 successor.

700(H) Squadron disbanded on 15 September 1967 and the first Front Line Squadron to receive Wessex HAS Mk 3 was 814 Squadron — six in October 1967. 706 Squadron, the pilot Advanced Flying Training Squadron at R.N.A.S. Culdrose, gradually exchanged its Wessex HAS Mk 1 for the improved version during 1967/68. 737 Squadron the Observer Advanced and Operational Flying Training Squadron at R.N.A.S. Portland received seven Wessex HAS Mk 3, 820 Squadron exchanged six Wessex HAS Mk 1's for four of the new Wessex prior to embarking on the newly converted helicopter cruiser, H.M.S. BLAKE. 819 Squadron, now based in Northern Ireland at R.A.F. Bally Kelly, (R.N.A.S. Eglinton had closed), re-equipped with four Wessex HAS Mk 3 and finally in October 1968 826 Squadron re-equipped with six aircraft destined for H.M.S. EAGLE.

The Flights on board the County Class destroyers converted to the new aircraft as they disembarked to R.N.A.S. PORTLAND for leave and maintenance.

The Wessex HAS Mk 3 continues to give good service to the Fleet in County Class destroyer flights and in 737 Squadron which is the Advanced Flying Training squadron for Observers and Sonar Operators. The rear cabin equipment of the Wessex HAS Mk 3 is very similar to that of its successor, the Sea King HAS Mk 1, and therefore backseat crew training is still effectively carried out using the older aircraft. The last Wessex HAS Mk 3 phased out from Front Line service (apart from County Class Flights) on 29 January 1971 when 819 Squadron disbanded at R.A.F. Bally Kelly. The aircraft had a very quiet and inauspicious career in the Front Line. It provided a nucleus of crews trained in the use of the new sonar and radar equipment in preparation for the introduction of the Fleet Air Arm's next generation of A.S.W. helicopters–with greater endurance, weapon carrying capability and more versatility in all roles from carrying underslung loads to Search and Rescue. The Wessex HAS Mk 3 introduced an entirely new technique to the Fleet, that of Helicopter In-Flight Refuelling (H.I.F.R). The

▲ Early days — **Whirlwind HAS 7**

Vhirlwind mk 9 — Modified for use in the antarctic flying from HMS Endurance ▼

▲ An early model — **Wessex HAS 1**

Wessex HV 5 — Helps out during flood relief in Pakistan ▼

Wessex HAS 3 — The Camel

▲ HMS Ark Royal SAR flight — **Wessex Mk 1**

▼ Above the clouds — **Wessex HV5**

Wasp HAS 1 in training colours ▼

The old — the **Lynx HAS 2** says good-bye to the Frigate HMS Lynx as she enters retirement

▲**Wasp HAS 1** fires an AS12 missile
The New — The **Lynx** takes off from HMS Birmingham ▼

A Gazelle of the Sharks aerobatic display team

▲**Sea King HAS 2** — over an angry sea

Sea King HAS 1 — carries a load of Landrover & trailer for The Royal Marines▼

technique was simplicity itself. The aircraft came alongside a ship, lowered its winch wire to which the ship attached a refuelling hose. Once winched up into the aircraft, the hose was connected to the aircraft fuel system and fuel taken on board. Crew changes were also possible whilst refuelling. The use of this technique came about as ship and helicopter anti-submarine screens extended out from the convoy or major unit being screened due to improvements in surface sonar and in submarine performance and weapons. A quick refuelling from the nearest surface ship was far more preferable than a 20 mile transit back to the parent ship, refuel and then transit back again to the screen. Limitations to the aircraft endurance have now become aircrew fatigue and maintenance constraints on hours flown. H.I.F.R. continues to be used wideiy in the Fleet Air Arm and it is especially useful where two aircraft are operating from a single aircraft deck such as that on an 'O' Class Tanker. In the event that the deck becomes 'foul', (that is when an aircraft has just landed and is unable to fold its blades,) the ship still has the ability to supply fuel to a second aircraft so that it can fly to another deck or wait until the 'foul' deck is cleared.

A Wessex Mk 1 of 771 Squadron practising Diver Dropping techniques in Mounts Bay, Cornwall.

With the introduction of the Wessex HAS Mk 3 and its successor the Sea King HAS Mk 1, the Wessex HAS Mk 1 was not retired from service. It continued to serve in 771 Squadron, the Search and Rescue Aircrewman Training squadron based at R.N.A.S. Culdrose until 1979. Up until December 1978 two Wessex Mk 1 served on board H.M.S. ARK ROYAL as her Search and Rescue Flight. 771 Squadron had recommissioned at R.N.A.S. Portland on 23 June 1967 with ten Whirlwind HAS Mk 7, these Whirlwinds were replaced with the Wessex from 1970 onwards and then in July 1974 with the formation of 772 Squadron, 771 Squadron had moved to R.N.A.S. Culdrose.

Only the Wessex HU5 remains in front line service in the Fleet Air Arm, the Wessex 3 and 1 having been overtaken by the advance of technology. Like the Whirlwind the anti-submarine versions of the Wessex proved tactics for the next generation of anti-submarine helicopters in a more than competent way.

CHAPTER 7

Small Ships Flights

It had always been the intention of the Royal Navy to get helicopters to sea in every available vessel. This can be seen from the initial trials in the S.S. DAGHESTAN in the 1940's right through to the present day when the aim has almost been achieved.

Early aircraft did not lend themselves well to being operationally viable as they lacked a good payload of equipment or weapons. Also the maintenance effort required would have been out of proportion to the flying achieved. Weather was another factor; helicopter flying at sea took a long time to mature as money was just not available to develop the techniques and technology required. Data collection however did begin in September 1946 when, as already mentioned, an R4B had landed on board H.M.S. HELMSDALE.

The first 'small ship' to be fitted with a platform from which helicopters would regularly operate was the survey ship H.M.S. VIDAL. Initial trials were carried out in 1954 and H.M.S. VIDAL then sailed for her survey task. The flight consisted of one Dragonfly Mk 3 and this aircraft was used in the annexation of the island of Rockall in 1955. The success of this venture led to the decision to fit a flight deck on H.M.S. PROTECTOR the Antarctic survey ship. In July 1955 H.M.S. PROTECTOR'S flight formed with two Whirlwind H.A.R. Mk 1 aircraft.

In 1957 the Royal Aircraft Establishment at Bedford started looking seriously into the problems of small helicopters landing and taking off from frigate sized ships. At this stage, in spite of all the trials, deck landings and take-offs that had taken place, there was surprisingly little concrete information available. However, data was forthcoming during 1957 when a Fairey Ultra Light helicopter carried out Deck Landing Trials on board H.M.S. GRENVILLE, a wartime frigate, in the Channel. During this trial wind speeds of up to 62 knots, a ship roll of 14 degrees and deck movement of 12 feet were experienced and data recorded. The Royal Navy was very pleased with the success and records from this limited trial.

However, this was all well in the future when the design work on the aircraft which was to be the forerunner of the Fleet Air Arm's first true small ship anti-submarine helicopter was begun in 1945. This aircraft was the W.14 Skeeter 1 built by CIEVRA. The first prototype flew on 8 October 1948 and the Ministry of Supply took an active interest in the progress of flying trials. Further prototypes improved performance and it was at this stage that the Royal Navy became interested and the Skeeter 4 was slightly redesigned to meet the Royal Navy small ship requirement and called the Skeeter 5. This work was actually completed by Saunders-Roe Ltd as this company took over CIEVRA in 1951.

The Skeeter 5 did not come up to expectations but Saunders-Roe were now aware of the Naval requirements and continued to work to produce a viable naval helicopter. Work was slowed up as the Army had accepted the Skeeter 4 and resources were not available to work on two separate designs at one time.

Eventually, as a private venture, Saunders-Roe produced an aircraft which they thought could replace the Army Skeeter 4 and also meet the Naval requirement. This aircraft was known as the Saunders Roe P.531. Many components in the P.531 were the same as used in the Skeeter, the major difference between the two aircraft being the power plant. The P.531 was powered by a turbine engine, the Blackburn built Turmo 603 s.h.p. engine derated to 325 s.h.p. The first prototype flew on 20 July 1958, the second shortly afterwards.

Westland Aircraft Ltd took over Saunders Roe in 1959 and continued work on the P.531. They completed two more prototypes up-rating the engine power using a Blackburn Nimbus engine (derated to 635 s.h.p.) and giving the aircraft a

A prototype Wasp, the P531 fitted with suction pad landing gear.

skid undercarriage. The first of these prototypes flew on 9 August 1959 and the second (powered by another engine, the Gnome H.1000, derated to 685 s.h.p.) flying on 3 May 1960.

In 1959 the Royal Navy carried out extensive flight trials on

The P531 fitted with skid landing gear landing on a Light Fleet Carrier alongside in Portsmouth Harbour.

the second of the prototypes built by Saunders Roe (with skids fitted), and on two further P531s designated P531-O/N. These second two aircraft were fitted with a long stroke quadricycle wheel undercarriage as well as landing skids.

In November and December of 1959 in a continuation of the trials the P531 successfully completed a series of Deck landings off Portland on board H.M.S. UNDAUNTED. H.M.S. UNDAUNTED was another wartime frigate conversion which had had a landing platform built above the Quarterdeck.

It was now obvious that the P.531 was to be the aircraft to equip the small ships flights of the 1960's. Some minor modifications such as tail folding so that the aircraft would fit into the frigates were still required. A major problem still not resolved was the form the undercarriage would take. In order to resolve the problem the Royal Aircraft Establishment at Bedford constructed a 'Rolling Platform' on the Bedford airfield.

The undercarriage problem was in fact twofold. Firstly the helicopter had to land without damage within predetermined limits. Secondly, once the helicopter was on deck the ship had to have the ability to manoeuvre tactically. If, during this manoeuvring, the undercarriage collapsed due to ship motion, any advantage gained from the landing would be worthless. So deck securing methods and undercarriage design were to be equally studied. Other operations such as weapon loading were also involved as a helicopter with no weapons is as useless as no helicopter at all.

The Bedford 'Rolling Platform' was therefore to assist in the research into all these matters. The platform is 24 feet square and rocks on a trolley which in turn oscillates to and fro along a track. The coordination of these motions simulate the motion of the deck of a frigate. By varying various 'cranks', degrees of roll between +/-2° and +/-15° can be simulated. The frequency of motion is altered by motor speeds and gear ratios. The track runs from inside a hangar, from which helicopter mock-ups can be suspended for stationary experiments. When live helicopter landings are programmed the 'Rolling Platform' runs up and down outside the hangar. Originally extensions were hinged to both ends of the platform to prevent the helicopter from sliding off and overbalancing, this modification has now been removed.

After consideration of many proposed Deck Securing Systems, two main ones emerged which appeared to meet the requirements. The first, Suction Pads on the aircraft, the second a Harpoon and Grid system.

The Suction Pad system revolved around having suction pads attached to the aircraft. Equipment in the aircraft would provide a negative air pressure for landing and a positive air pressure for moving the aircraft on deck, somewhat like a hovercraft. Trials were carried out in H.M.S. UNDAUNTED in 1961 and in H.M.S. ASHANTI in 1962 using this system. Difficulties were encountered, with pad sealing, the roughness of the deck and differential pressures on the pads when moving around the deck being the main contributory factors. However it did appear that the system could be made to work.

The Harpoon and Grid system was never tested at sea, only at R.A.E. Bedford in 1960. It consisted of a harpoon attached to the underside of the fuselage which engaged a grid on the deck. A Dragonfly carried out the flying trials whilst a dynamic full scale P531 model fuselage was trialled on the Rolling Platform in the hangar. Technical difficulties and weight considerations caused this idea to be dropped. This system however reappears again on the Lynk HAS Mk 2 which is covered later in this chapter.

In the event all investigations came to nought as the decision was taken to fit the P.531 with a four point castored wheel undercarriage, lower the minimum collective stop to decrease the residual lift on the rotor to just less than 10% of the aircraft weight, give the Flight Deck a high coefficient of friction and finally to fit a centre-mounted swivel-link under the fuselage which could be released by the pilot at the moment of take-off. The combination of all the above allowed the pilot to swivel into wind using the tail rotor with the wheels set tangentially on a circle. For shore operations the front wheels were to be set fore and aft. Special nylon lashings for quick attachment and release were developed at R.A.E. Bedford. The first of class trials on all these pieces of equipment were successfully completed on board H.M.S. LEANDER in 1963.

In 1961 the Royal Navy placed its first substantial order for the production of the Wasp HAS Mk 1 as the aircraft had now become known. The final number ordered was 96. On 4 June 1963 the Wasp Intensive Flying Trials Unit, 700(W) Squadron, formed at R.N.A.S. Culdrose and the M.A.T.C.H. system was born. M.A.T.C.H. stands for Medium-Range Anti-Submarine Torpedo Carrying Helicopter and gave surface ships, carrying a Wasp helicopter, the ability to deliver one or two homing torpedoes on top of submarine contacts obtained on their Medium-Range sonar sets which were outside the range of the

An 829 Squadron Wasp from RNAS Portland with a full weapon load over HMS AURORA.

ships own anti-submarine weapons. Provided other ships in the force were capable of radar control of the Wasp then they could 'borrow' the Wasp for their own attacks.

On 4 March 1964, 700(W) Squadron was disbanded, having successfully evaluated the aircraft, and in its place 829 Squadron formed. 829 Squadron was to parent the Wasp Flights and the County Class Wessex Flights which have already been mentioned, to provide Operational Flying Training for Wasp pilots who had completed the Advanced Flying Training with the three Wasps of 706 Squadron, to formulate new Wasp tactics, and to undertake trials.

On 1st December 1964 829 Squadron moved to R.N.A.S Portland, a more logical base than R.N.A.S. Culdrose due to the proximity of the Naval Base which was responsible for the working-up of ships. This now allowed the ships Flight and the ship to work-up together. Following the move 771 Squadron was disbanded and its Whirlwind HAS Mk 7 aircraft were transferred to 829 Squadron. 829 Squadron now had additional tasks to those already mentioned. Using the Whirlwind HAS Mk 7s it trained Search and Rescue crews and also became the Fleet Requirements Squadron for Portland providing aircraft for everything from VIP transport to Gunnery Radar Alignment.

Between November 1961 and April 1964 seven Tribal Class frigates (Type 81) commissioned, between March 1963 and February 1973 twenty-six Leander Class frigates commissioned, between 1966 and 1972 nine Whitby Class (Type 12) frigates were converted to take a Wasp Flight and were re-classed as Rothesay Class and three Hecla Class survey ships were commissioned between December 1965 and May 1966. The fourth Hecla Class, H.M.S. HERALD did not commission until 31 Oct 1974. These ships gave a maximum of 46 decks from which the Wasp could operate on a permanent basis. On shore 829 Squadron had 7 training aircraft, Dartmouth Air Experience Flight had one aircraft and 3 training aircraft were attached to 706 Squadron at R.N.A.S. Culdrose for Advanced Flying Training. In fact the build-up of Wasp capable ships and aircraft numbers just about kept pace with each other, but there were occasions when ships went to sea without a Flight embarked. As ships started to return for Long Refits the situation eased and by the late 1960's the Flight System run by 829 Squadron had the situation completely under control. As an experiment 848 Squadron was allocated a Wasp for liaison duties but this was discontinued after two years as the benefits were outweighed by the maintenance of the aircraft and aircrew cross-training.

The versatility of the Wasp HAS Mk 1 was increased on 13 February 1969 when the AS-12 missile was first fired from the aircraft. This missile, an air to surface wire-guided weapon, was designed to combat the growing threat from the Soviet missile armed fast patrol boats. Small ships at last had an 'over the horizon' capability for engaging enemy targets.

In 1970 the responsibility for parenting the County Class Wessex HAS Mk 3 Flights was handed over to 737 Squadron. 829 Squadron had no shore-based Wessex HAS MK3's and the task of parenting these Flights was starting to overload an already hardworked Headquarters Staff. 737 Squadron was operating Wessex HAS Mk 3 aircraft for Observer and Aircrewmen Advanced and Operational Flying Training and the move was a logical one. Three years earlier, on 23 June 1967, the Whirlwind HAS Mk 7 aircraft had been returned to 771 Squadron when it had reformed at RNAS Portland as the Fleet Requirements Unit and for Basic Aircrewmen training.

The final split for 829 Squadron came on 27 January 1972 when 703 Squadron formed at R.N.A.S. Portland tasked with the Advanced and Operational Flying Training for all Wasp aircrew. 703 Squadron took over the seven aircraft from 829

Squadron and later on the three aircraft from 706 Squadron. The formation of 703 Squadron was the last act in the rationalisation of Wasp training and operating. All Wasp activities were now concentrated at R.N.A.S. Portland, by now well on the way to becoming Europe's largest helicopter operating base.

829 Squadron was, and still is, the largest helicopter squadron in the Fleet Air Arm with over 40 aircraft. Its task is not likely to diminish, indeed, in the next few paragraphs it will be seen that the task could well increase.

The Wasp had its disadvantages and the technology of the 1950's was not capable of overcoming such limitations as lack of payload and endurance. However as the 1950's ended the Royal Navy and Westland Aircraft Limited were already working on new helicopter designs. By 1964 the concept of the rigid rotor and a helicopter to which it could be fitted had developed, and the British Army was also taking an interest. This design was called the WG13. The Royal Navy displayed great caution at first over the new design, and it was not until 1966, when a naval variant of the WG13, the WG13V, was put forward for consideration that any real interest was shown. Working in parallel in France, Sud Aviation was looking at a similar sized helicopter for both naval and army use. The British and French Governments realising the need to cut design and development costs decided to collaborate in the development, not only of the WG13 both also another new light helicopter, the Gazelle. The agreement was signed by both Governments in 1967. (A further helicopter, the Puma, was also involved but this aircraft was never considered for Naval service.)

By the end of 1967 the go-ahead for the building of four naval prototypes had been given (plus eleven others for other trials). However only three Naval prototypes were built (plus nine others) due to the cancellation of other variants.

The first WG13 prototype flew on 21 March 1971. This was not a Naval prototype, the first of those flew on 25 May 1972. The new aircraft had been named the Lynx and the naval version was to be known as the Lynx H.A.S. MK2 and a total of 60 were ordered as an initial buy.

Flight trials indicated a few problems such as vibration, engine over-heating and excessive oil consumption. These problems were in the main overcome by the time the first production Naval aircraft was wheeled out in December 1975.

On 17 September 1976 the Lynx Intensive Flying Trials Unit,

700(L) Squadron, was formed at R.N.A.S. Yeovilton tasked with evaluating the Lynx H.A.S. MK2 in all its roles. 700(L) Squadron was in fact unique in Fleet Air Arm history as it was the first ever fully integrated Anglo-Dutch helicopter squadron with both nationalities maintaining and flying each others aircraft. This had arisen as the Netherlands Navy had bought the Search and Rescue version of the Lynx, the H.A.R. 25, and the advantages to both services in running a dual I.F.T.U. close to the Westland Helicopter Limited factory at Yeovil were obvious. The Dutch portion of the I.F.T.U. departed for Holland on 13 May 1977 well satisfied with their new aircraft.

In February 1977 the I.F.T.U. carried out Deck Landing trials on H.M.S. BIRMINGHAM in the Channel and at around the same time started trials in the tactical use of the Lynx HAS Mk 2.

The Lynx has the ability to carry out all the functions of the aircraft it replaces but with an improved endurance and payload. The aircraft is fitted with a highly sophisticated

A Lynx HAS 2 preparing for an approach to the deck of HMS ARDENT.

electronic emmission detector which is essential in the highly electronic war at sea against ships and submarines. Twin engined reliability gives the crew of one pilot and one observer a greater margin of safety over the sea. Lastly, the introduction of the British Aerospace Corporation missile, the Sea Skua early in the 1980's will give the Lynx an Anti-Ship capability equivalent to that of a frigate. This missile, which will sea-skim in its approach to the target, is carried on wing pods, two to either side.

700(L) Squadron disbanded on 16 December 1977 and on 3 January 1978 reformed as 702 Squadron the Lynx Training and Flight Parenting Squadron still at R.N.A.S. Yeovilton. At a future date the parenting of the Lynx Flights will be assumed by 829 Squadron at R.N.A.S. Portland with the training remaining with 702 Squadron much the same as 703 Squadron performs the Wasp training task at present.

The first operational Lynx Flight was that of H.M.S. SIRIUS, a Leander Class frigate, which formed over Christmas 1977. Several new Classes of ships have been and are being built to take the Lynx to sea, The Amazon Class (Type 21) being the first, H.M.S. AMAZON was commissioned on 11 May 1974, seven further ships of this class followed. The Sheffield Class (Type 42) came next with H.M.S. SHEFFIELD commissioning on 16 February 1975. Thirteen more of this class will eventually join the Fleet. Finally the Broadsword Class (Type 22) of which the first two are at sea. Sixteen of this class of ship are expected to be ordered. Most Leanders and other frigates will be modified to take the Lynx as time and money permits. Again the situation of out of phase ships and aircraft has meant that some Lynx capable ships have had to forego their Flights, mainly the Amazon Class. Ships without a Lynx are being allocated a Wasp Flight for the time being.

One of the major modifications required by an older ship embarking a Lynx is the fitting of a grid to the centre of the Deck Landing Area. This is a part of the French developed tethering system used by the Lynx. This system called Harpoon was fitted in an Alouette in 1961 and trialled at R.A.E. Bedford, however it was not used until the Lynx came into service. Harpoon consists of a hydraulically operated retractable harpoon deck-lock mounted in the underside of the Lynx. In the deck is mounted a honeycomb grid into which the harpoon engages to provide a positive deck lock. The Lynx undercarriage is a tricycle arrangement with the rear two wheel castoring through 27 degrees and the front undercarriage

Directors onboard RFA ENGADINE prepare to receive two Lynx HAS 2.

assembly castoring through 90 degrees. This allows the aircraft to swivel about the harpoon into the relative wind. Should the pilot attempt to take-off with Harpoon engaged the deck lock has a weak link designed to part at 8000 lbs strain. This system is now well proven and will hold the Lynx stable at up to +/− 35 degrees of roll before toppling.

The Royal Navy will continue to operate the Wasp and the Lynx together for a few years to come. The Wasp proved that single small helicopter flight embarked in a ship was a viable proposition, the Lynx HAS Mk 2 has now made a quantum jump with the introduction of new equipment and missiles. The Lynx HAS Mk 2 will go a long way to making up for the deficit of surface ships that the Royal Navy permanently suffers.

Sea King HAS 1 aircraft of 820 Squadron overflying their parent ship, HMS BLAKE.

CHAPTER 8

The Kings of Anti-Submarine Warfare

The Royal Navy had progressed with its sonar carrying aircraft through the Whirlwind HAS Mk 7, which had a short endurance and no weapon carrying capacity when sonar was carried; to the Wessex HAS Mk 1, which had twice the endurance of the Whirlwind HAS Mk 7 but when fitted with a weapon had a drastically reduced endurance, and also had a sonar which was rapidly becoming obsolete; to the Wessex HAS Mk 3, which improved the anti-submarine capability by adding radar and a vastly improved sonar, but brought down the endurance yet again and did not have a practical weapon carrying capability. Somehow the new radar and sonar, plus weapons and endurance, had to be combined into one aircraft. Advances in submarine technology, especially in the missile field, were making it imperative to be able to send an A.S.W. helicopter up to 100 miles from a screened force, remain on task for at least an hour and then return to the force.

Such an aircraft had been flying in the United States since 11 March 1959 when the XHSS-2 made its maiden flight. The production HSS-2's had been serving with the U.S. Navy as SH-3A's since September 1961. This aircraft would fit into all existing British aircraft carriers, including the two converted helicopter cruisers, H.M.S. BLAKE and H.M.S. TIGER, and provided the price was right, would meet the requirements for an A.S.W. helicopter for the 1970's and possibly the 1980's.

In 1967 Westland Helicopters Ltd acquired a licence from Sikorsky to build a version of the HSS-2 in Britain. This aircraft was to be called the Sea King HAS Mk 1 and the Royal Navy placed an order for 60 aircraft valued at £24 million. This price included the fitting of British electronics and engines, plus the fitting of the radar and sonar equipment used in the Wessex HAS Mk 3.

On 7 May 1969 the first Westland built production Sea King made its maiden flight at Yeovil and on 19 August 1969 the Sea King HAS Mk 1 Intensive Flying Trials Unit, 700(S) Squadron, formed at R.N.A.S. Culdrose with the task of evaluating the aircraft prior to entry into service. The aim was to achieve 2,700 hours of flying time with 6 aircraft between 19 August 1969 and 30 April 1970 when the unit would disband. 1800 hours were specifically concerned with evaluating the anti-submarine role of the aircraft. The I.F.T.U. completed the task with time to spare and came to the conclusion that the Sea King HAS Mk 1 was the most efficient integrated hunter/killer weapon system yet to be designed. Ten years later that conclusion is still valid.

The Sea King HAS Mk 1 package consisted of firstly the Sonar Type 195, the Ekco radar and the Marconi Doppler Navigation system as already tried and thoroughly tested in the Wessex HAS Mk 3. The airframe, built at Westlands was to the American design with the addition of the radar radome aft of the rotor head. The major changes to the American design were all hidden within the body of the aircraft. The 1400 s.h.p. T58-GE-10 American engines were replaced by two 1500 s.h.p. Gnome H1400 engines. All electronics were British, one of the major advances being in the Automatic Flight Control System, the Mk 31, which continued to be supplied by Louis Newmark. The reliability of this type of system had been so improved that the system returned to a Simplex operation, that is, with no fail-safe operation. The partially "Observer operated" facility was removed, proving no loss. Since its introduction into service the A.F.C.S. Mk 31 has proved itself reliable and well capable of operating up to its design limits.

The greatest advantage over all other British A.S.W. helicopters was what had been sought for many years, endurance and payload. The endurance was now in the region of four hours in an A.S.W. profile with a transit endurance of up to five hours. Fitted with 2 torpedos the Sea King HAS Mk 1 could still complete a four hour sortie in temperate climates and for the loss of about an hours fuel a total of 4 torpedos could be carried. The Sea King HAS Mk 1 has been likened to an A.S.W. frigate and indeed whilst airborne on the screen ahead of a force the Sea King is the equivalent of a frigate, in some respects better if unpredictability of movement and rapid investigation of long range radar and sonar contacts are taken into account.

On 24 February 1970 the first front line Sea King squadron, 824 Squadron, was formed at R.N.A.S. Culdrose, bound for H.M.S. ARK ROYAL. 824 Squadron had seven aircraft, one more than its normal complement. The seventh aircraft replaced the C.O.D. (Carrier Onboard Delivery) Gannet which was normally used for mail and stores runs to and from shore. As this extra aircraft retained its A.S.W. equipment it was also used to supplement the squadron's A.S.W. task. This aircraft was withdrawn in 1975 when there was a shortage of Sea Kings in both Front Line and Training Squadrons.

In 1970, 706 Squadron, the Advanced Flying Training Squadron at R.N.A.S. Culdrose started to receive its Sea Kings for pilot training. 737 Squadron, the Observer and Aircrewman Training Squadron at R.N.A.S. Portland did not receive its aircraft until 1973 as Front Line Squadrons had priority.

The next Front Line Squadron to receive Sea Kings was 826 Squadron in June 1970 when it exchanged 6 Wessex HAS Mk 3 for 6 Sea King HAS Mk 1 and embarked onboard H.M.S. EAGLE for the carriers final deployment. In 1972 H.M.S. EAGLE went into reserve and 826 Squadron reduced to 4 Sea Kings HAS Mk 1 in preparation for embarking in H.M.S. TIGER the second helicopter cruiser.

819 Squadron had disbanded on 29 January 1971 when R.A.F. Bally Kelly had closed. However there was a pressing requirement for an A.S.W. helicopter squadron to be based close to the Clyde submarine exercise areas and Prestwick, the civil airport close to Glasgow, was chosen as the operating base. Former U.S.A.F. buildings were taken over and on 9

February 1972 819 Squadron commissioned at R.N.A.S. Culdrose with 6 Sea King HAS Mk 1 moving up to Prestwick in October of the same year. 819 Squadron also assumed, in addition to their A.S.W. tasks, the Department of Trade's Search and Rescue requirement for Western Scotland.

820 Squadron received 4 Sea King HAS Mk 1 in exchange for its 4 Wessex HAS Mk 3 in December 1972. 814 Squadron reformed at Prestwick on 30 March 1973 with 4 Sea King HAS Mk 1, having disbanded back in July 1970 when H.M.S. HERMES was taken into dockyard hands for her conversion from a fixed wing to helicopter carrier. 814 Squadron was destined to pioneer a major change in A.S.W. thinking within the next few years.

By the end of 1977 Sea King HAS Mk 1 aircraft were employed by the following Front Line units, 814 Squadron (H.M.S. HERMES), 819 Squadron (PRESTWICK), 820 Squadron (H.M.S. BLAKE), 824 Squadron (H.M.S. ARK ROYAL) and in 826 Squadron (H.M.S. TIGER). In the Training role it was solely in service in 706 Squadron at R.N.A.S. Culdrose. It had served briefly with 737 Squadron at R.N.A.S. Portland between 1973 and 1975 but the pressure from Front Line squadrons wanting Sea Kings had caused its withdrawal and it was replaced by the Wessex HAS Mk 3.

A Sea King HAS 2 of 824 Squadron carrying its full load of torpedoes.

The Royal Navy ordered a further 21 aircraft in 1975 to bring the total order to 81. These additional aircraft were to be designated Sea King HAS Mk 2. The two modifications introduced on this new aircraft were the fitting of an extra blade to the tail rotor (giving a total of 6), giving the pilot more sideways control, and an increase to the fuel jettison rate so that fuel could be jettisoned more quickly. At the same time, although not part of the Mk 2 aircraft solely, a number of anti-icing modifications were fitted so as to improve the performance of the aircraft in icing or near icing conditions. The most obvious of these anti-icing modifications was the fitting of a large 'barn-door' in front of the engine intakes above the cockpit. All Sea Kings have now been modified to Mk 2 standard, the conversions being carried out by R.N.A.Y. Fleetlands at Gosport.

In 1975 814 Squadron was tasked with carrying out a feasibility trial on passive sonar equipment. Four aircraft were converted to take "Jezebel" equipment, and to carry sonobuoys. By dropping passive sonobuoys into the water and monitoring the sound collected by their sensitive hydrophones it is possible, by analysis and triangulation, to establish the position of certain submarines at far greater ranges than with active sonar.

In an attempt to reduce the cabin temperature in Sea Kings a trial aircraft was painted white in the areas most affected.

The trial indicated that Sea King helicopters equipped with this equipment were a viable proposition and the decision was taken to equip all 814 Squadron aircraft with Jezebel in order to build up a nucleus of crew experience in passive sonar operation. Jezebel was rapidly becoming outdated and design studies were started on new systems for fitting in the Sea King replacement (see next chapter). Jezebel fitted aircraft retain their active sonar and therefore are able to swiftly 'localise' and attack any Jezebel contacts obtained.

The success of the Jezebel Sea King indicates that, before the Sea King replacement is introduced into service, all Sea Kings will be fitted with passive sonar equipment and that the progression to more up to date and advanced passive sonar is moving fast.

The Sea King has proved to be a reliable and efficient A.S.W. aircraft and proof of this is shown by the fact that the Royal Australian Navy, the Indian Navy, the Pakistan Navy and the Eygptian Armed Forces have all brought A.S.W. Sea Kings. The Belgian Air Force, German Navy and Norwegian Air Force have bought the Search and Rescue version as have the Royal Air Force. All sales of the aircraft have included training for the country's aircrew which has been carried out at R.N.A.S. Culdrose by Fleet Air Arm aircrew. A special Foreign Training Unit was set up and trained each countries aircrew on their own version of the Sea King. This training benefited considerably from the Sea King Simulator, a full scale Sea King cockpit mounted on a motion platform, located at R.N.A.S. Culdrose. The rear cabin is also simulated but not located on the motion platform. The cockpit and rear cabin can be linked

A Sea King HAS 1 from RAE Boscombe Down on loan to RAE Pershore for Lynx radar trials.

electronically for full-crew training or both can be operated independently.

The success of the Sea King Simulator, which commenced operation in 1970, can be gauged by the fact that Norwegian, and Royal Air Force crews return to R.N.A.S. Culdrose for continuation and emergency training in the Simulator. The German Navy and Royal Australian Navy have bought their own simulators having been impressed by the benefits to be gained by such an equipment. The Danes although not operating the Westland built Sea King, but the Sikorsky version, also send their Search and Rescue crews for emergency training.

Future Naval helicopters will all be supported by a training Simulator, the Lynx HAS 2 Simulator is being built at present and as the Sea King Replacement develops, the Simulator for the aircraft will progress.

Simulators have an established role in the training of Fleet Air Arm helicopter crews. They have replaced initial flying sorties by a cheaper and more cost effective method of learning basic skills. They have also replaced more advanced flying sorties where the demonstration of (say) an emergency recovery technique could lead to a potentially dangerous situation. The student now has an 'aircraft' which he can 'crash' without loss to the Navy.

The Sea King story has by no means finished and A.S.W. Sea Kings will be serving with the Royal Navy for a very long time yet. The Sea King has made possible the introduction of passive sonar detection equipment into the helicopter field whilst still retaining the active sonar and weapons in the air, a very important advance in the A.S.W. battle.

Helicopter flying is not without its accidents. This Sea King HAS 1 was successfully recovered from the sea 20 miles off Portland Bill.

A Hiller HTE2 of 705 Squadron at RNAS Lee on Solent.

CHAPTER 9

The Way Ahead

The preceding eight chapters have taken the helicopter up to its present advanced state. This final chapter looks at planned future developments.

Firstly, a look at the Training Squadrons. 705 Squadron, the Basic Pilot Training Squadron which has served the Fleet Air Arm continuously for 32 years. In 1975 the Squadron was re-

A Hiller 12E of 705 Squadron at RNAS Culdrose.

equipped with Westland Gazelles replacing the aged Whirlwind HAS 7's and Hiller 12E's. (The Hiller, in two versions, the HTE2 and 12E, had served with 705 Squadron since 1953.) The Gazelle was part of the Anglo French agreement signed in 1967 and its introduction enabled the Fleet Air Arm to start the student helicopter pilot on a turbine powered aircraft thereby reducing training time on conversion to more complicated operational aircraft. 705 Squadron will remain in service as long as helicopters fly with the Fleet Air Arm.

The two major Operational and Advanced Flying Training Squadrons, 706 Squadron for Anti Submarine Warfare (with the Sea King Mk 2) and 707 Squadron for Commando Support (with the Wessex HU 5) will continue at R.N.A.S. Culdrose and R.N.A.S. Yeovilton respectively.

737 Squadron at R.N.A.S. Portland which provides Basic and Operational Flying Training for Observers and Aircrewmen and also parents the County Class Wessex Mk 3 Flights, (only two of which are still operational) is training an increasing number of students using 14 Wessex Mk 3. A major part of Operational Flying Training for Sea King aircrew was transferred from 737 Squadron in February 1979. Final Operational Flying Training for Sea Kings Pilots, Observers and Aircrewmen now takes place in 706 Squadron. This entailed an increase to 13 Sea King and further increases are likely due to additions to Operational Flying Training Courses. It is of interest that 706 Squadron also parent two Royal Air Force Sea King Mk 3. These are flown and

Three Gazelle aircraft of 705 Squadron airborne from RNAS Culdrose. These aircraft form part of the 'Sharks' display formation team.

maintained by RAF personnel providing training for RAF Sea King Mk 3 crews.

As already stated elsewhere the Wasp still has a long future and 703 Squadron at R.N.A.S. Portland will continue to provide the Basic and Conversion training required. 702 Squadron will provide similar services for Lynx aircrew. In time 702 Squadron will presumably move to R.N.A.S. Portland and take over from 703 Squadron. 829 Squadron will eventually re-equip with Lynx but this has not yet started and all parenting of Lynx Flights is being done by 702 Squadron. The complete conversion from Lynx to Wasp will take many years to complete and 829 Squadron will remain the Fleet Air Arm's largest Squadron.

772 Squadron, the Fleet Requirements and Search and Rescue unit at R.N.A.S. Portland with Wessex HU 5 will remain unchanged. So will 771 Squadron at R.N.A.S. Culdrose which provides Basic Training for Aircrewmen and Search and Rescue facilities again with Wessex HU 5. Conversion of 771 Squadron from Wessex Mk 1 to Wessex HU 5 took place during the early part of 1979 and the last Wessex Mk 1 in Naval Service left R.N.A.S. Culdrose on 9 August 1979. It was fitting that this very aircraft had also been the last one to leave H.M.S. ARK ROYAL as she entered Plymouth Sound for the last time in December 1978. The build-up of Wessex HU 5 in 771/772 Squadrons has been met by taking the aircraft from the Commando Squadrons prior to their receipt of Sea King Mk 4 aircraft.

The Front Line Commando Squadrons, 845 and 846 Squadrons are at present without permanent sea-going bases but they are still firmly committed to supporting the Royal Marine Commando's on NATO's Northern Flank. Both Squadrons now have 12 Wessex HU 5. In early 1980, 846 Squadron was reinforced by four Sea King Mk 4 to lift the increasingly heavier loads demanded by the Commando Forces. Four Sea King Mk 4 will be added to 845 Squadron in 1981. The first Sea King Mk 4 arrived at R.N.A.S. Culdrose from Westlands on 3 December 1979 for checks and painting prior to issue to 846 Squadron. A permanent feature for many years to come will be the CLOCKWORK series of winter exercises based in Northern Norway. These exercises are designed to maintain a nucleus of Air and Ground crews experienced in Artic Operations and Warfare. Although the Commando Squadrons no longer have their permanent bases at sea, they retain the ability to deploy at short notice on any Royal Navy or Royal Fleet Auxiliary ship capable of carrying and operating

The first Sea King HU 4 arriving at RNAS Culdrose for pre-issue checks.

helicopters. Frequent exercises maintain this important capability.

1978/1979 saw major changes for most of the Front Line Anti Submarine Squadrons. 826 Squadron disembarked from H.M.S. TIGER for the last time in the summer of 1978. H.M.S. TIGER then went into Reserve. 826 Squadron remained shore-based until H.M.S. BULWARK emerged from a refit in Portsmouth Dockyard in March 1979. H.M.S. BULWARK is now fully modified to carry A.S.W. Sea Kings and is filling the gap between the "paying off" of H.M.S. ARK ROYAL, H.M.S. TIGER and H.M.S. BLAKE and the acceptance into service of H.M.S. INVINCIBLE in the early 1980's.

With the final paying-off of H.M.S. ARK ROYAL in December 1978, 824 Squadron became the Headquarters Squadron for Sea Kings operating from Royal Fleet Auxiliaries including a new Class of RFA, the FORT Class. This class comprises, at present, two ships — the FORT GRANGE and FORT AUSTIN. Each ship is capable of operating up to four Sea Kings during exercises or in wartime, however the normal complement is one Sea King for Vertical Replenishment duties. 824 Squadron has increased its aircraft complement from six to eight aircraft. For convenience of operation the Squadron has split into four Flights of two aircraft, A,B,C and D Flights.

820 Squadron served onboard H.M.S. BLAKE with four Sea Kings until 5 December 1979 when they disembarked to R.N.A.S. Culdrose. H.M.S. BLAKE continued on to Portsmouth and into Reserve.

814 Squadron continues to pioneer work with Jezebel operating from H.M.S. HERMES who will continue on through the 1980's. H.M.S. HERMES will undergo a refit during 1980/1 to enable her to carry and operate the Sea Harrier.

At Prestwick, 819 Squadron with six Sea Kings provides a useful service to the Clyde Submarine Base and to the Scottish Search and Rescue facilities. The Squadrons future at Prestwick appears assured.

Late in 1978 the Ministry of Defence announced that Britain, France and Italy would be cooperating in a new helicopter for the 1980's, the WG 34. Design studies have already been in progress for a number of years and reports indicate that the new helicopter will be in the 24,000 to 28,000 lb All Up Weight range. It will be capable of All Weather operation, single pilot operation and will carry the latest homing torpedoes, electronics, radar and sonics. The WG 34 will eventually replace the Sea King but the two will probably operate side by side for a number of years. An update of the present Sea King HAS 2 is therefore projected. Firstly to enable the WG 34 and Sea King to work together and secondly to update the existing radar and sonics to a standard required to cope with faster and deeper submarines. This updated aircraft will be known as the Sea King HAS 5 and should be in Front Line service in 1981.

In summary, it is safe to say that the importance of the helicopter, be it Anti Submarine or Commando, Missile or Torpedo Carrier, or just Search and Rescue, to the Royal Navy is at a peak. Never before has the helicopter possessed so many options which are available to a Force Commander. Unlike other nations the Royal Navy has made it policy that helicopter aircrew and in particular Anti Submarine aircrew, are trained to make decisions on the spot. This is particularly important in times of radio silence. This policy plus the training and equipment provided makes the Fleet Air Arm helicopter force one to be reckoned with. There is no reason to suppose that the situation will change in the foreseeable future.

CHAPTER 10

Picture Library

A Whirlwind HAS 22 landing onboard an unknown ship in the early 1950's. The Flight Deck Officers method of communication is of particular note!

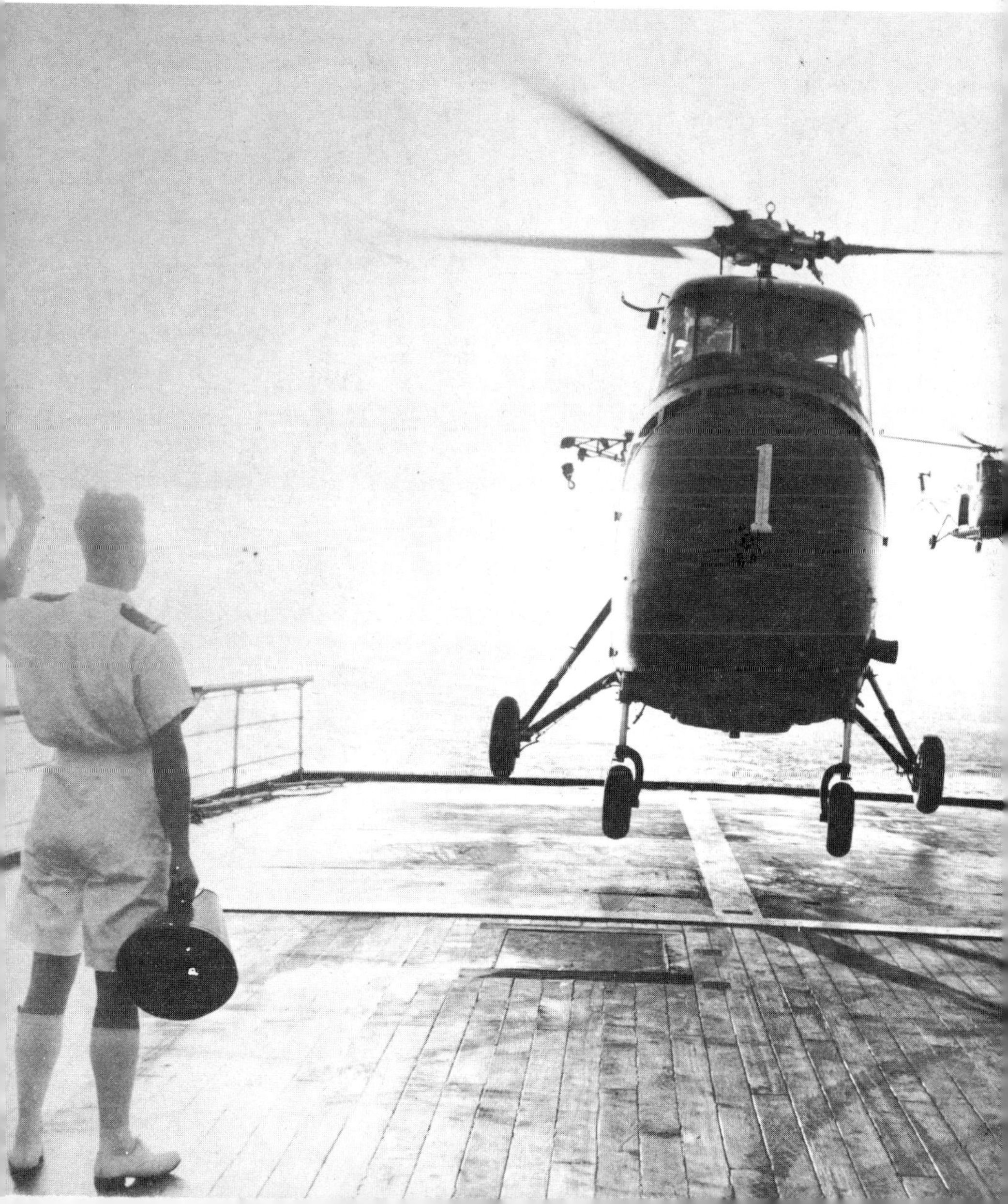

Whirlwind Mk 7 of 820 Squadron from HMS ALBION showing the Flag over New Zealand.

Whirlwind Mk 7 of 820 Squadron onboard HMS ALBION involved in an unusual Flight Deck accident.

A Whirlwind HAS 22 from HMS PROTECTOR operating in Antartica.

Wessex HAS Mk 1 of 819 Squadron operating from HMS ARK ROYAL.

Roping Drills being practised by a Wessex HAS Mk 1 of 814 Squadron from HMS Victorious in Hong Kong.

Wasp lashed down on the minute Flight Deck of HMS NUBIAN.

Wasp of 829 Squadron at RNAS Portland firing an AS 12 missile.

HOT...A scene at a Forward Operating Base at Mersing, Malaysia. A Wessex HU 5 of 848 Squadron lifts a Haflinger transport.

COLD...A Wessex HU 5 from HMS HERMES operating near Bardufoss in Norway during Exercise CLOCKWORK.

A 771 Squadron Wessex Mk 1 practising Cliff Rescue techniques somewhere in Cornwall.

An 845 Squadron Wessex HU 5 on patrol over Maalla Straight, Aden.

Sea King HAS 1 of 824 Squadron lifting an unserviceable Wasp from HMS ARROW to RNAS CULDROSE for repairs.

A Sea King HAS 1 of 706 Squadron acting as a flying crane. The underslung Skyraider was being transferred to RFA ENGADINE from RNAS Culdrose enroute to the Fleet Air Arm Museum.

A Sea King HAS 2 of 824 Squadron escorting a Wessex Mk 1 of the ARK ROYAL SAR Flight back to ARK ROYAL. The Wessex Mk 1 has an under-slung load of a practice torpedo.

Heavy lift techniques being practised by a Sea King HAS 1 of 700(S) Squadron

An action shot which appeared in BBC TV's 'Sailor'. A Sea King HAS 1 of 824 Squadron rescuing a sick American sailor in a stretcher who had been swept from the USS BERGAL during a casualty evacuation flight in mid Atlantic. Petty Officer Aircrewman WITHALL received a bravery award for diving in and assisting the American sailor.

RFA FORT GRANGE with one of its two Sea King HAS 2 from 824 Squadron.

Sea King HAS 1 of 700 Squadron transfers personnel to the Polaris submarine HMS RESOLUTION

APPENDIX A. PRINCIPLE DETAILS OF FLEET AIR ARM HELICOPTERS.

DETAIL / AIRCRAFT	POWER PLANT(S)	MAIN ROTOR DIAMETER	FUSELAGE LENGTH	HEIGHT	MAXIMUM TAKE OFF WEIGHT (lbs)
SIKORSKY R4B (HOVERFLY 1)	180 h.p. WARNER SUPER SCARAB R-550-1. 7 cylinder radial.	38′ 0″	35′ 5″	12′ 5″	2535
SIKORSKY R6A (HOVERFLY 2)	245 h.p. FRANKLYN	38′ 0″	47′ 11″		2600
WESTLAND DRAGONFLY HR1, HR3, HR5	550 h.p. ALVIS LEONIDES 50. 9 cylinder radial.	49′ 0″	40′ 10″	12′ 11½″	5870
SIKORSKY HRS-2 (H.A.R. 21)	550 h.p. PRATT & WHITNEY R-1340-57	53′ 0″	42′ 2″	13′ 4″	6835
SIKORSKY H04S-3 (H.A.S. 22)	700 h.p. WRIGHT R-1300-3	53′ 0″	42′ 2″	13′ 4″	7500
WESTLAND WHIRLWIND H.A.R. MK.1 (Series 1)	600 h.p. PRATT & WHITNEY WASP R-1340-40	53′ 0″	41′ 8½″	13′ 2½″	7500
WESTLAND WHIRLWIND H.A.R. MK.3 (Series 1)	700 h.p. WRIGHT CYCLONE R-1300-3	53′ 0″	41′ 8½″	13′ 2½″	7500
WESTLAND WHIRLWIND H.A.S. MK.7 (Series 2)	750 h.p. (derated) ALVIS LEONIDES MAJOR 755/1	53′ 0″	41′ 8½″	13′ 2½″	7800
HILLER H.T.E.2	210 h.p. FRANKLYN	35′ 0″	38′ 8″	9′ 6″	2500
WESTLAND WESSEX H.A.S. MK.1	1450 s.h.p. NAPIER GAZELLE 161	56′ 0″	48′ 4½″	14′ 5″	12600
WESTLAND WASP H.A.S. MK.1	968 s.h.p. (derated to 685 s.h.p.) Rolls Royce NIMBUS 101	32′ 3″	30′ 4″	8′ 10″	5300
WESTLAND WHIRLWIND H.A.R. MK.9 (Series 3)	1050 s.h.p. ROLLS-ROYCE GNOME H1000	53′ 0″	44′ 2″	13′ 2½″	8000
HILLER 12E	305 h.p. LYCOMING VO-540	35′ 0″	28′ 0″	10′ 1½″	2700
WESTLAND WESSEX H.A.S. MK.3	1600 s.h.p. NAPIER GAZELLE 165	56′ 0″	48′ 4½″	14′ 5″	12600
WESTLAND WESSEX HU MK.5	2 x 1250 s.h.p. ROLLS ROYCE GNOME	56′ 0″	48′ 4½″	14′ 5″	13500
WESTLAND SEA KING H.A.S. MK.1/MK.2	2x1500 shp ROLLS ROYCE GNOME H1400	62′ 0″	55′ 9¾″	15′ 6″	20,500 (MK1) 21,400 (MK2)
WESTLAND GAZELLE	590 shp TURBOMECA ASTAZOU III	34′ 5½″	31′ 2¾″	9′ 0¼″	3747
WESTLAND LYNX H.A.S. MK.2	2 x 830 s.h.p. ROLLS ROYCE BS 360-07-26 GEM.	42′ 0″	38′ 3¼″	11′ 3″	8000
WESTLAND SEA KING HU MK.4	2 x 1500 s.h.p. ROLLS ROYCE GNOME H-1400-1	62′ 0″	55′ 9¾″	16′ 6″	2100

NOTES: GPMG — Cabin Mounted General Purpose Machine Gun FFFMG — Fixed Forward Firing Machine

CEILING (feet)	TYPICAL RANGE (miles)	PRIMARY ROLE(S)	CREW	ARMAMENT	EQUIPMENT	
00	220	Fleet Requirements & Pilot Training	1 Pilot			
00		Fleet Requirements & Pilot Training	1 Pilot			
00	300	Search and Rescue & Fleet Requirements	1 Pilot 1 Aircrewman		Winch	Day SAR only
00	440	Commando	1 Pilot 1 Aircrewman		Winch	
00	400	Commando & Anti-Submarine	1 Pilot 1 Observer 1 Aircrewman		Winch/Sonar	Day ASW only
00	320	Search and Rescue	1 Pilot 1 Aircrewman		Winch	Day SAR only
00	320	Search and Rescue	1 Pilot 1 Aircrewman		Winch	Day SAR only
00	335	Anti-Submarine & Commando	1 Pilot 1 Observer 1 Aircrewman	G.P.M.G. — CDO 1 Torpedo — ASW	Winch/Sonar	Day ASW only
00	135	Pilot Training	1 Pilot			
00	390	Anti-Submarine & Commando	2 Pilots 1 Observer 1 Aircrewman	G.P.M.G. & FFFMG — CDO + SS 11 2 Torpedoes — ASW	Winch/Sonar	All weather ASW and SAR. 1 Pilot operation by day
00	315	Anti-Submarine & Anti-Surface	1 Pilot 1 Aircrewman	2 Torpedoes or 2 AS 12	Winch	
00	300	Search and Rescue	1 Pilot 1 Aircrewman		Winch	Day SAR only
0	205	Pilot Training	1 Pilot			
00	390	Anti-Submarine	2 Pilots 1 Observer 1 Aircrewman	2 Torpedoes	Winch, Sonar Radar	All weather ASW and SAR. 1 Pilot operation by day
0	478	Commando & Anti-Surface	2 Pilots 1 Aircrewman	SS 11/FFFMG/AS12 2″ Rockets 2 Torpedoes	Winch	1 Pilot operation by day
0	598	Anti-Submarine	2 Pilots 1 Observer 1 Aircrewman	4 Torpedoes G.P.M.G.	Winch, Sonar Radar	All weather ASW and SAR
0	403	Pilot Training	1 Pilot		Winch	
0	489	Anti-Surface & Anti-Submarine	1 Pilot 1 Observer	4 Sea Skua Missiles 2 Torpedoes	Winch Radar	All weather
0	598	Commando	2 Pilots 1 Aircrewman	G.P.M.G. 2″ Rockets	Winch	All weather

APPENDIX B

Dunking Sonar

The principle on which active sonar works is exactly the same as a person shouting in a large empty canyon hears his echo. Sound is produced at a point source, on reaching an obstacle the sound is reflected and some of this reflected sound returns to the point source.

Sound at the normal human frequency range does not travel in water and so high frequency sound is used instead. This sound is produced by electrical means in the 'transducer', the part of the sonar set which is lowered into the water from the helicopter. On meeting a solid body in the water some of the sound will be returned to the transducer which acts not only as a transmitter but also as a receiver. The transducer is designed in such a fashion that the sound is sent out in a pre-determined beam shape, i.e. a pencil beam like a searchlight or over any area desired. Designing the beam shape, taking into account the amount of power available to push out the sound is a critical factor in the design of a successful sonar set.

As the speed of sound in water is known, the range of the object reflecting the sound back to the transducer is easily worked out. In early helicopter sonar sets the range was displayed on a moving roll of electrically sensitive paper, the increase of sound in the returns boosting the electrical signal. In modern sonar sets the information is displayed on Cathode Ray Tubes.

In order to obtain the bearing of the reflecting object it is necessary to allow the transducer to rotate freely about its cable and introduce an input from the helicopter's compass system. The Observer in the helicopter now has a position of the 'contact' in range and bearing from the helicopter's position. The bearing can be displayed on the Cathode Ray Tube or an on simple pointer.

The above has not taken into account two factors which affect the detection of submarines to a greater extent than any other. Firstly, sound is affected in its passage through the water by water temperature, salinity, density and depth (pressure). Secondly, there are many objects in the water, some natural, others man-made, which reflect sound and give a definite impression of being a submarine, but which in fact are not. The subject of sound in water is a science in itself and it is not intended to go deeply into the problems, merely to illustrate them with some examples.

Isothermal water is water where there is no increase in temperature with depth and in general sound travels in a straight line path in isothermal water. Unfortunately there are few areas in the oceans where these conditions prevail to any great depth before coming to a 'layer'. At the layer the temperature normally drops and continues to drop until it reaches 2,500 metres where it stabilises at about 4°C. In some shallow seas like the Mediterranean it stablises at about 15°C. On passing through the layer the sound is refracted downwards, however in some cases, dependant on the depth of the layer, the temperature difference and the angle at which the sound hits the layer, the sound may not penetrate at all and may even be reflected. Sometimes a 'sound duct' may be formed with a layer of isothermal water in between two layers of increasing or decreasing temperature. Transmitting inside this 'duct' may produce long ranges on any contact inside the duct, but short ranges on those outside.

It is in the use of water conditions that the helicopter excels. A surface ship sonar is limited to using its sonar a fixed distance below the surface of the water, although a Variable Depth Sonar (V.D.S.) for ships is in use with varying success. A helicopter with its variable transducer depth can exploit the water conditions and use the depth best suited to obtaining long ranges. Of course the submarine is also an expert at using the water conditions and the battle between helicopter and submarine is very much 'cat and mouse' with the unit making the wrong decision the loser.

The second factor mentioned that the helicopter crew has to take into account is the number of natural and man-made phenomena in the ocean which purport to be submarines. Animal life such as whales and large shoals of fish are examples of natural phenomena. They produce an 'echo' of the right proportions and in most cases they will have movement. In certain parts of the world, such as off the Norwegian coast, the mixing of fresh and salt water can produce very good echoes but these do tend to be of short duration. Underwater disturbances and waves provide echoes which can give the most experienced crews some nasty moments. Man-made echoes produced are 'knuckles' in a ships wake when it turns and likewise from a turning submarine. Bubbles of air fired from submarine torpedo tubes which are designed to lure the helicopter crew onto a false contact are frequently encountered. In shallow waters there are wrecks and rocks

HMS HERMES being screened by a Sea King HAS 1 of 814 Squadron. Note the sonar being lowered.

which give very good returns especially if a tide is running at the time.

As can be seen from the preceding few paragraphs the science of anti-submarine warfare using dunking sonar is a complicated one. The Royal Navy has developed it well and is without doubt a world leader in the field.

There is however a second type of sonar — passive sonar, which does not rely on transmitting into the water, rather it listens to sounds in the water which are being radiated from the submarine. This type of sonar has been mentioned in Chapter 8 when referring to 814 Squadron carrying out Jezebel trials.

Passive sonar work consists of dropping a series of buoys in a pre-determined pattern into the water. Each buoy deploys a very sensitive microphone (hydrophone) beneath it to a preset depth. Sound information is transmitted back from the buoys to the helicopter by radio. By analysis of the sound and by triangulation of the sound strengths it is possible to 'fix' a submarine's position relative to the buoys.

The advantages of passive over active sonar are twofold. Firstly, in certain conditions and on certain submarines the passive system can obtain longer ranges than active and sometimes identify the type of submarine detected. Secondly, the submarine commander does not know the buoys have been laid as they are silent. One disadvantage of present buoys is that although they can detect earlier and at longer ranges than active sonar, the actual 'localisation' of the contact using the buoys can take a long time and dropping a weapon in the water on passive information alone is not an accurate method of attack. Hence 814 Squadron aircraft are retaining their active sonar to complement the passive system.

Research and development on improved passive and passive/active systems is being carried out with the object of installing the most up to date system in the Sea King Replacement.

APPENDIX C

Guide to present day Naval Air Stations and their helicopters.

The object of this guide is to give aircraft enthusiasts and others an idea of where they may see Naval Helicopters.

There are three principal Naval Air Stations and three other locations where Naval Helicopters may be seen on a regular basis.

Principal Naval Air Stations

R.N.A.S. YEOVILTON (Ilchester, Somerset)

702 Squadron	LYNX HAS 2	Pilot and Observer Training Parents all Lynx Flights.
707 Squadron	WESSEX HU 5	Advanced Flying Training.
845 Squadron	WESSEX HU 5	Front Line Commando Squadron
846 Squadron	WESSEX HU 5 and SEA KING HU 4	Front Line Commando Squadron

The FLAG OFFICER NAVAL AIR COMMAND (FONAC) who is in charge of all Naval Air Stations and shore based aircraft has his offices at YEOVILTON.

The FLEET AIR ARM MUSEUM is adjacent to the airfield and well worth a visit.

R.N.A.S. PORTLAND (Portland, Dorset)

703 Squadron	WASP HAS 1	Advanced Flying Training.
737 Squadron	WESSEX HAS 3	Advanced and Operational Flying Training. Parents Wessex 3 Flights.
772 Squadron	WESSEX HU 5	Search and Rescue. Fleet Requirements.
829 Squadron	WASP HAS 1	Parents all Wasp Flights.

R.N.A.S. CULDROSE (Helston, Cornwall)

705 Squadron	GAZELLE	Basic Pilot Training.
706 Squadron	SEA KING HAS 2	Advanced and Operational Flying Training. Long Range Search and Rescue.
	SEA KING HAR 3	Parents RAF Training Unit.
771 Squadron	WESSEX HU 5	Basic Flying Training. Search and Rescue. Fleet Requirements.
814 Squadron	SEA KING HAS 2	Front Line Squadron
820 Squadron	SEA KING HAS 2	Front Line Squadron
824 Squadron	SEA KING HAS 2	Front Line Squadron
826 Squadron	SEA KING HAS 2	Front Line Squadron

Other locations

PRESTWICK AIRPORT (Ayr, Ayrshire)

819 Squadron	SEA KING HAS 2	Front Line Squadron. Search and Rescue.

R.N.A.S. LEE-ON-SOLENT (Lee-on-Solent, Hampshire)

781 Squadron	WESSEX HU 5	Search and Rescue. VIP Transport.

R.N.A.Y. FLEETLANDS (Gosport, Hampshire)

The Royal Navy Aircraft Yard Fleetlands undertakes helicopter repair and modification work for all three British Services. A variety of helicopters will be seen flying in and out of the airfield.

ACKNOWLEDGEMENTS

Without the help of those listed below, this book would not have been possible:

Commander H.M.A. HAYES, RN (Ret'd)
Commander B. SARGINSON, RN
Commander R. WARD, RN
Commander G.N.I. HARVEY, RN

Lieutenant Commander C. WHITE, RNVR
Lieutenant Commander D. FOLEY, RCN (Ret'd)
Lieutenant Commander P.H. STUART, RN (Ret'd)
Lieutenant Commander M. MADDOX, RN
Lieutenant Commander D.A. BLYTHE, RN
Lieutenant Commander P.J.W. STEVENS, RN
Lieutenant Commander P.S. BUCKLEY, RN
Lieutenant Commander S. TURTON, RN
Lieutenant Commander R.A.Y. BRIDGES, RN
Lieutenant Commander C.P. YOUNG, RN
Lieutenant Commander N. HENNELL, RN
Lieutenant J.E.K. SANDERSON, RN (Ret'd)
Lieutenant J. BROWN, RN (Ret'd) Naval Historian.
Lieutentant R. SAFE, RN

Third Officer B. STONE, WRNS

Chief Petty Officer (Photographic) S.R. HOBDEN

Mr Elfan ap Rees, Westland Helicopters Limited. Historian.

Mr J.B.B. Johnston RAE BEDFORD
Miss M. Dyke RAE BEFORD

The Photographic Sections DPR(N) — Ministry of Defence
HMS EXCELLENT
HMS ARK ROYAL
RNAS CULDROSE
RNAS YEOVILTON
Westland Helicopters Ltd.